SATAN HAS NO POWER OVER YOU

OVER YOU

You Belong to Jesus Now

"Scripture taken from the New King James Version. Copyright ©
1982 by Thomas Nelson, Inc.

ISBN:978-1-950252-06-0

SATAN HAS NO POWER OVER YOU

OVER YOU

You Belong to Jesus Now

By Summer McClellan

Other books by Summer McClellan

The Impossible Marriage

Grace What Is It?

Faith What Is It?

Passing the Tests of Life

To *Lon and Barb DeNeff*

Table of Contents

"I am the light of the world. He who follows Me shall not walk in darkness but have the light of life."   John 8:12

Introduction

He has delivered us from the power of darkness and translated us into the kingdom of the Son of His love.
Colossians 1:13

"Satan has no power over you! He has no power over you, no power over you!" My sister Carol's voice was booming in my ears. I had never heard her voice carry so much authority; it was like there was thunder in it. My mouth dropped open in amazement as she continued to prophesy the same words over and over again to me.

"Satan has no power over you, no power over you; he has no power over you." I felt as if I was seated before God, and it was Him speaking to me. God's words were like a battering ram pounding against the lies in my mind. I was afraid of the devil. I'd suffered from oppression for years. I was afraid of getting the devil angry with me....yet... I know God doesn't lie. There was a war in my mind. I was trying to believe the words I was hearing. God continued speaking the same words through my sister over and over until a glimmer of light began to shine through.

Could it be true?

The title of this book *Satan Has No Power Over You, You Belong to Jesus Now,* that is my goal for you. As you read this book, to hear it, to believe it, to know it, to bring it to pass in your life. This is the key to supernatural living, to living in your godly inheritance now. You have been born into the kingdom of Heaven, there is no poverty no sickness, no oppression there.

I suffered through extreme oppression for many years. I didn't know back then we could get free of the devil. I thought we were stuck down here with the devil while God is up there in Heaven. While I believed that I stayed oppressed. {We are seated with Him in heavenly places Ephes. 1:6} As the light of God's truth broke through, I began to get free in areas. I began to live supernaturally in areas as I realized Satan had no power over me because I belong to Jesus.

CHAPTER ONE

Darkness

For you were once darkness, Ephesians 5:8

I came to the end of myself at the age of fourteen. It is hard to believe someone so young could come to the end of themselves, but I did. I didn't want to live anymore; I was beyond miserable. I would drink anything I could get my hands on or take hands full of pills out of the medicine chest or if I couldn't get anything else, I would sniff anything with fumes.

I have always felt afraid and empty, as long as I can remember. My twin sister Carol and I were born to a single teenage mother; life was unpredictable. Our mom had to work, and we were left with an assortment of different people.

We would cry for our mother and that would get us

yelled at. I hated to be yelled at. It left me terrified. I would have never ever done anything on purpose to get myself yelled at. When I did get yelled at, I would cry and that would get us yelled at more.

We were left to ourselves a lot. Our poor mom when she was home, she would sleep, she was tired. We would wander around by ourselves, my twin sister and I, and amuse ourselves. One-time mom woke up and found we had painted ourselves green with oil paint. We did it as a surprise for her, boy, was she surprised. Another time she woke up to find us riding our tricycles down the middle of the street. Then there was the time we found some cloth in the attic and cut it up to make our dolls some new clothes. Mom was very angry, I cringed as she yelled at us, it had seemed like such a good idea. And like the time we climbed up to the top cupboard to eat all the yummy orange baby aspirin, that seemed like a good idea too.

Our poor mother was so overwhelmed, and she would yell. Yelling confused and terrified me. I was never trying to be naughty. It seemed the grown-up world was unpredictable and getting yelled at was inevitable. You never knew when it would happen, it just always happened.

I felt very little and very afraid. And I couldn't trust adults. There was no one to go to for help. They just didn't listen. One babysitter would molest us or maybe torture us would be a better word. She would take us in the bathroom and take our clothes off us, and jam things up us, like a toothbrush. We were only about three or four. I remember getting away from her and going from one neighbor to another knocking on their doors, screaming and crying and begging for help. I didn't get any.

I begged my mother not to leave us with her again. "She sticks stuff up us" I said.

"You won't have her much longer." mom said.

I remember giving up inside, I just gave up.

But things did get better when I was seven my mom married a wonderful man. He adopted my sister and me.

"It's over" I told myself and it was.

Our mother is very trusting of everyone and has never been a good judge of people's character. She would trust anyone, but our new dad had a keen sense of judgment. We were never left with anyone questionable again.

Still, I was void. It seemed like there was nothing on the inside of me. I had no feelings and no personality. I whined and complained a lot. I always said I had a headache or stomachache.

My happiest memories were the times spent with my sister playing with our dolls. My doll was Snooks, and my sister's doll was Mugsy. As far back as I can remember Carol and I played with our dolls. They were our world, our own world. We would laugh until we cried; our dolls had so much personality, extreme personalities, something I was void of. It was a world that belonged to me and Carol and Snooks and Mugsy. We played with our dolls every day and slept with them every night. Snooks and Mugsy could say the things we couldn't say and do the things we couldn't do. They were our way of coping.

I didn't know much about God. I don't remember grown-ups explaining God to me. I remember a Catholic girl that lived down the street telling me Jesus was a little man in your heart that made a black mark on your heart every time you sinned or told a lie. Of course, I believed her. I used to wonder if my skin was going to turn black if Jesus ran out of room making black marks on my heart. I was always checking my arms for black marks.

We were faithful church attenders of the world's deadest church. They didn't preach the gospel. My Jesus figure was Santa Claus; he was someone who loved me enough to bring me presents. When I found out he wasn't real I was devastated. I then wondered if God was real too or just something else grown-ups made up.

I was always very compliant at home or at school, it may have been my fear of getting yelled at. I just couldn't deal with that; I would be reduced to tears and unable to speak. Then came my preteen years, I was vulnerable because I was so empty inside. I thought of myself as so worthless that I didn't think anything mattered. I had no brakes. Once I started sliding the wrong direction in my life I went quickly. There was nothing inside me to stop myself or slow myself down.

I began smoking about twelve years old with some kids in the neighborhood. All of a sudden, I was in; I was a part of a crowd. I suddenly had friends because I smoked, it got me noticed.

Satan laid a trap, and I stepped right in. I almost got out though, after a month of smoking my dad caught me.

He grounded me to a chair for a month. I actually felt relief, "That's over" I said to myself. I didn't like making my parents mad, but after a month I wasn't grounded anymore. The trap was still out there waiting for me, I stepped back in again.

It was almost like getting "unsaved." Something was filling the void. I was so empty I progressed quickly into darkness. The devil had me and he knew it.

The area we lived in had a real demonic stronghold over the youth. Drugs were a real problem at our schools in junior high and the high school. There were a lot of drugs and everything that went with it. It seemed like the majority of the kids were into it. Later when we moved to a different area, I couldn't believe the difference, most kids didn't drink and smoke and do drugs.

One thing I also became void of was shame. My shame wire was disconnected, remember nothing mattered. The more trouble you got into in our neighborhood it seemed the cooler you were. Bad was good and good was bad. I lacked any ability to hold my own and I was empty and hungry and an easy target for the enemy.

I progressed in a short time from cigarettes to swearing and lying, drinking, drugs, stealing, sex and the occult. I noticed other kids had inner brakes, they dabbled in what the rest of us were doing but it didn't consume them. But as for me I was empty, and nothing mattered. I didn't care that I was destroying my own soul.

I'll give you an example of how much I, Summer, didn't matter. One time when I was thirteen years old, I went with a girlfriend and her parents to a baseball game in Detroit. I don't know why I went I have never liked sports. While I was there, I wandered around bored. A tough gang of boys saw me alone and surrounded me and took me into the parking lot. I was raped by every one of them, I couldn't even say how many times. I never even told anyone, I just told myself it didn't matter, nothing mattered.

The problem is it does matter, somewhere inside of me a person was living, I didn't know her, I couldn't feel her, but she was being destroyed. Now I know thirty-nine years later how much it does matter, everything matters, everything we do, every word we say, every day we live, every person we encounter. It all matters. Every beat of your heart and every breath you take, every hair on your head, every care, every hurt, and every tear they all matter. I matter and you matter; your soul matters because it is precious to Jesus. He gave everything He has to save it. Even Satan is jealous for it and fights for our souls because he wants to destroy them. EVERYTHING MATTERS!!!!

Another thing I noticed in my couple of years of deep darkness was I had an unholy protection. Satan would keep me from getting caught. {I needed to get caught.} I remember one night all of us kids that hung out together broke into some cabins in a nearby camp to have a drinking party. I got up and left and a few minutes later the police came, and they all got in trouble. There were many incidents like that.

I have noticed that Satan does this, this unholy

protection. He does this with people sold out to him and in his trap, like rapists and murderers. It is not because he cares about them but because it does the most harm. It keeps them progressing in evil and it keeps them hurting others. It takes prayer to break his unholy protection.

At the same time, Satan was destroying me, I had gotten involved in the occult and strange things started happening to me. I first noticed I would lose things, like my hairbrush. I always kept my hairbrush in the drawer in the bathroom. I opened the drawer, one morning, to use it and it wasn't there. I looked all over for it, then I opened up the drawer again and it was right there. Then it started getting to be bigger things, like my lamp at my bedside, I tried to turn it on, one night and it was gone and so was the bedside stand. I went and told my mom, and we came back in the room, and it was there. Doors would open and close and I would hear knocking. I was very troubled at this time, and I would sleepwalk. I would wake up in all different places. It would scare me. Satan thought he had me. I was trapped in sin and in torment and I was miserable.

Satan's big worry about me was my dad. My dad fought for me by punishing me. He'd ground me or whip me with a belt. If I had any brakes at all it was my dad. And then there was Satan's biggest threat of all, my sister Carol, she got saved.

CHAPTER TWO
Light

"To open their eyes and turn them from darkness to light, and from the power of Satan to God, that they may receive forgiveness of sins and an inheritance of those who are sanctified by faith in Me".   Acts 26:18

We had moved to our new neighborhood when I was eleven. We had been driving a long distance to church every Sunday to go to the old church, the dead church. After a few years of this my parents decided to find a church in our new area because of the drive. We began visiting churches.

We tried a new church that was meeting in a school, they didn't have a building yet. It was a nondenominational church. To us it was radical, it was unbelievably radical. They preached the gospel, they had an altar call, and they baptized people. They would actually dunk them under water. We had never heard of such things. It was an archaic church compared to churches we later joined. The sermon was about salvation every week. They didn't believe in healing or tongues or any of the fun stuff, but at the time to us they were radical.

My sister wanted to go forward and be saved but she didn't have the nerve; instead, she prayed with Billy Graham on television. Then she got baptized at church.

She started telling me about the Lord. I didn't want to hear it. It would make me cringe. It was light and I was darkness. I would swear at her. She also began to pray. Her prayers were effective; the first thing that happened after she started praying was the unholy protection was broken by her prayers. I started getting in trouble, a lot.

I had a babysitting job one night; after I put the kids to bed, I used the house for a party. My dad found out and whipped me with a belt. I am grateful for a dad that whipped me. I believe in this kind of punishment when all else fails, because it worked, it produced fruit in my life. It reconnected my shame wiring; all of a sudden, I was totally ashamed of myself. Bad didn't seem so good anymore, it seemed bad. I got in trouble again; I got caught shoplifting. My mom and dad were so fed up with me they barely even spoke with me.

God was doing something in me, I was becoming miserable. I felt trapped in misery. God was opening my eyes, too, though at the time I didn't realize it was God. He was opening my eyes to how little I really meant to all these friends I had. I didn't really mean anything to them.

God was using my sister Carol's prayers and He was carefully orchestrating my rescue from darkness. The last straw had to do with our dolls Snooks and Mugsy. God had to break one last bond because it was enough to keep me from Him.

There was a boy I liked a whole lot; he was always in trouble and my parents didn't want him around. One day when mom and dad were gone, and I had let this boy into the house, he did the unthinkable. He saw Snooks and Mugsy and he picked them up and wrote on their faces with an ink pen.

I was horrified and disgusted. I don't think anyone, but my sister could possibly realize what Snooks and Mugsy meant to me. I quickly tried to wipe the ink from their faces, it wiped off Snooks, but it wouldn't come off of Mugsy. I didn't know what to do. I knew Carol would never forgive me and she would hate me forever. I hid Mugsy. This pushed me over the edge. I loathed what I had let happen to Mugsy. My sister was the only person that didn't seem disgusted with me. I needed her.

"Where's Mugsy?" My sister wanted her doll. Everyone in the house was looking for Mugsy. I was trying everything to get that ink off her face; she had ink across her eyebrows and a moustache.

After several days I couldn't stand it anymore. I brought Mugsy to Carol and told her what happened. She didn't hate me. She showed me love. God used that too. It made me cling to Carol. I was so miserable I started following my sister around.

I had come to the end of myself. I wanted to kill myself. I had been thinking about it for days. I was scared so I finally decided I wouldn't live much longer anyway; my lifestyle was too destructive.

All of a sudden, I was sick of darkness and Carol was light; she was shining Jesus' light. I slept on the floor next to my sister's bed. She kept up with her constant telling me about Jesus. I wasn't swearing anymore when I heard His name. I was too desperate. I was listening to her.

She was sitting on her bed, and I was lying on her floor crying. I was so sick of me, and I was depressed. I don't know if words can express my misery. She kept telling me Jesus loved me.

Her bedroom door was closed but as she was speaking Jesus walked through her bedroom door. I didn't see Him with my eyes, but I felt Him. His presence was magnificent. He was more real than the things I could see. As He walked through her bedroom door, His majesty filled the room. Then He came and stood next to me on the floor and spoke to me.

"Summer" He said, "I don't care what you've done, I love you."

I sat up and yelled, "I want God!"

Instantly the extreme darkness that filled me was gone and I was filled with light. Misery was replaced with peace and joy.

I sat up amazed. "I've got God now," I kept saying. "I've got God now." I jumped up and ran into the living room where my dad was sitting in his chair and reading his paper. "I've got God now" I told him.

"Yeah right" he said, he didn't even look up from his paper.

He was pretty fed up with me. I didn't care, I was so happy. I didn't know a person could feel this happy. The world was a different place. I had never noticed the sky was so blue or the grass was so green or the flowers and the trees. Everything was beautiful and I had never noticed it before.

The next day I threw away all my records, Alice Cooper and Pink Floyd, there was a whole pile. I didn't want them. I ran down the street to tell my friends I had God now. I found them all together. I had a bag of pot hidden outside and they had found it and were smoking it. I started to get mad but realized I didn't care; I didn't need anything but God. "I've got God now" I told them.

I didn't know too much about God, but what I knew I liked. He was with me. He was with me on the bus to school. He was with me at school, and He was with me at home. I put Jesus stickers on my clarinet case and wore Jesus pins. I tried to tell people what happened to me. I'd say, "I've got God now." It wasn't very effective; I got a lot of blank stares. What everyone seemed to understand better was that I was going to be baptized.

"Summer!" I heard my name called. A group of loitering kids on the corner called me over. This was a week or so after my change. "I heard you're getting baptized," the boy that called me over said.

"Yes" I replied.

"I wish God would get a hold of me," he said in

front of everyone.

It just so happened when the day for my baptism came; the church baptistery was being repaired. I was very publicly baptized in our front yard, in the lake we lived on. A large crowd of people from our church stood in our yard singing hymns as I was baptized.

My mom and dad gave me a Living Bible for a baptism present. I kept my nose in it all the time. I had a lot to learn. As much as I had embraced darkness now, I was embracing the light. I wanted as much of God as I could get.

CHAPTER THREE

Oppressed

I will say to God my Rock, "Why have you forgotten me? Why do I go mourning because of the oppression of my enemy?" Psalm 42:9

I had been a Christian about six months when my grandmother died. I had grown as a Christian somewhat, but our church didn't have a lot of teaching. Most of what I learned I learned by reading my Bible, reading Christian books and the Lord speaking to my heart.

That night we had gotten a call about grandma; she had had a stroke and was in the hospital. It was a school night; I was in the ninth grade.

Something was wrong with me that night, not just being worried about grandma, something more. I felt agitated, as the night went on the agitation grew.

I remember being in our bathroom and I was holding my hand mirror in my hand; it was an expensive one I had gotten for Christmas from grandma. It was like the one she had on her dresser. I was holding my mirror when a wave of something horrible hit me, feelings like I had before I was saved. I felt confused and didn't know what was going on. I put my mirror down too harshly and it cracked. I couldn't believe I broke my beautiful mirror; I sat there in disbelief, staring at my broken mirror.

Then I heard a horrible voice in my head, but it sounded distant. "I am coming to get you and take you to hell." It said.

I went in the living room to be near my mom and sister. I was feeling miserable, and the agitation kept getting worse and the voice continued. Every time I heard it, it sounded closer.

"I am coming to get you and take you to hell."

I didn't have much teaching on the devil. I didn't know he was a fallen angel. I thought he was more like an anti-god, sort of like God only evil. I had no idea God was more powerful than he was. I felt doomed and began to cry. Soon I was writhing on the floor in agony. I could feel something evil coming closer.

My mother didn't know what was going on.

"What on earth is that matter with you?!" she shrieked, wondering why I was acting so strangely.

At that point, whatever it was, entered the room and I physically saw it. It was at my feet. Three things happened simultaneously. In sheer terror, I somehow propelled myself from lying on my back, on the floor near my mother's feet, to on her lap. Then I landed back on the floor facing the other direction.

I also screamed as I was in midflight, "Mom call a minister!"

And lastly, I heard Jesus audibly. I had no idea He was in the room until I heard His voice. From hearing where the sound of His voice came from, I realized He was standing next to me before I flew onto my mother's lap.

"Leave her alone she belongs to Me!" He demanded in that wonderful voice that is like no other. The tone of His voice was past firm, He sounded angry.

Whatever it was, that was tormenting me it fled His voice. It was immediately gone. I felt it getting farther away just as I had felt it drawing nearer.

I have to stop right here and tell you something. I am telling you with tears streaming down my face. Jesus is my hero. I love Him. He is my everything. I can't express in words how much this moment meant to me, over thirty-six years ago, when silently Jesus came next to me when I needed Him. I didn't know He was there until He spoke. He has promised to never leave us. I wonder how many times He is standing beside us when we need Him, and we don't even know He is there. Those words He spoke became one of my treasures. *I belong to Him*. Those words are like a diamond ring I wear on my finger, my treasure.

Sometime in the night that night my grandmother died. The night she died I started having trouble. I started having night terrors and nightmares. My mother always said grandma complained of terrible nightmares.

Two weeks after grandma died, my dad's father died. Then soon after my uncle died. Then a cousin committed suicide and another cousin was killed in a car accident. Then a friend of mine died. My family was going through spiritual warfare that seemed to center around me. We did not know anything about spiritual warfare, but we were going to learn.

I have learned something about the devil, if he can't get you, he will attack those around you. His evil, prideful, heart viciously seeks revenge.

This same year my dad took a new job in a new city. After we moved, we started attending a powerful full gospel church that taught about spiritual warfare.

One time my parents were hosting a prayer fellowship from the new church at our house. I was having trouble that night. I kept seeing horrible faces looking in the window at me and then they would be gone.

I started crying and everyone gathered around me to pray. One couple there knew a lot about spiritual warfare. They explained a lot to my parents. I noticed dad was really listening. They had me renounce any occult activity I had ever taken part in.

This is important because involving yourself in the occult gives Satan rights over you; even small things like reading your horoscope are an absolute no no.

I renounced the witchcraft I had gotten into in my darkness period; also, before we had gotten saved when we went to the dead church, we had a Ouija board. We had thought of it like a game, it is not a game, it is involvement into the occult and opens you up to demonic activity.

Our new church friends taught us how we have authority over the enemy in Jesus' name. We were learning. Many times, after prayer I would have a time of peace but then trouble would reoccur.

I was having a difficult period soon after. I was having frequent night terrors; it was like wrestling something in my sleep that was paralyzing me. During this time, I saw posters all over town for a miracle meeting. The poster said there would be miracles, healings and exorcisms.

I told my mom I wanted to go. The poster gave me hope. I just knew if I got to that meeting the Lord was going to do something for me.

Our pastor at church said not to go to the meeting, they had never heard of the speaker and thought he may be a fake.

I was determined to go, so my mother took me.

The man had rented the biggest hall in town. There were only a few rows taken, most of the huge auditorium stood empty.

After the service I went up for prayer and he prayed for me. All I told him was, "I have bad dreams."

This evangelist began to pray for me. He was bold. He said, "A witch has put a curse on you. There is a demon on your back, he does not possess you, he is oppressing you."

Then he rebuked it. I felt such power and I fell to the floor like a feather. This had never happened to me before. When I got up, I felt so light and free. Something had left me. I did handsprings all the way to the car. I wrote the man to thank him. I did much better for a while, although I was still afraid at night.

One night I felt afraid, so I slept in my parents' bed with them. {Everyone in my family was tired of this but they put up with me.} As I lay there, I was talking to the Lord instead of sleeping.

"Lord I just want to praise you" I said.

"Then go to the prayer meeting tonight with your mom and dad." I was surprised to hear the Lord answer so clearly. I had already planned to see a friend that night.

"I don't want to go to the prayer meeting, Lord" I told Him. "I am going to see my friend."

I wondered why He wanted me to change my plans. "I don't want you alone tonight." He answered my questioning heart. The prayer meetings lasted very late, and I would have been home way before my parents.

"Okay" I told the Lord.

As I got up that morning, I saw Jesus standing at the end of the hallway. He was beautiful; he looked a lot like Salman's Christ. I got so excited I kept running back and forth between the hall and mom and dad's room yelling "Mom, dad, Jesus is in the hall!" If He was going to say anything more, I missed it because I was so excited to see Him.

That night after we all went home and got to bed, I woke up. I saw eyes looking at me in the dark. I can only see a few inches without my glasses, so I put my face closer to see what I was looking at.

I screamed. I was looking into a demonic face of an

ape like creature. We were eyeball to eyeball. Before I could finish screaming dad was running down the hall yelling "In the name of Jesus."

Dad said the Lord had woken him up. Dad was ready for something to happen. The thing looked startled and turned and disappeared like a shadow into the crack between the bed and the wall.

The Lord had a plan for my deliverance again. We were having dinner after church that Sunday with some of my parent's friends from church. While we were there I started shaking uncontrollably; my parents filled them in on what I had been going through. They all gathered around me for a powerful time of prayer. Things were okay again for a while.

Soon I was having trouble again. My dad, having learned a lot really stepped in. My dad is not a person who is easily intimidated. He used to work for GMAC, a company that loaned money to people to buy cars. My dad's job was to repossess the cars that people didn't pay for. This is a tough job, and it takes a tough guy to do it. Sometimes dad would get threatened. This would terrify my sister and I; we didn't want our dad to get hurt. Dad never seemed scared; he had a saying when someone was after him. "The bigger they are the harder they fall," he would say.

Dad was the same way in the spiritual realm. He was not intimidated by the devil. Dad had gotten saved and baptized shortly after Carol and I had. After we moved, he became a part of Full Gospel Businessmen's Fellowship. It was a wonderful group that was leading men into the deeper things of God.

Sometimes God would wake dad up before I had trouble and dad would be waiting. Even though dad was brave, I was not. I had never stood up to anything my entire life, human or otherwise. My philosophy was to hide. After that first meeting that my mom had taken me to, where the speaker prayed for me, I was a little braver but not much. I was still easily terrified. It took me years to overcome fear.

Usually, I got in bed with my sister every night but if I were really scared, I got in my mom and dad's bed. I would get so scared; I would literally try to burrow myself under my dad.

Dad really prayed for me. One day he had a breakthrough. Dad was in his room praying for me. Suddenly he was standing next to himself and looking down on his body still knelt in prayer. Standing next to him was a large angel with a sword. The angel told dad he was there to help. Then he told dad there were demons in the house.

Even though this angel was there to help dad, it was dad who spoke to the demons, in the name of Jesus. The angel and dad went from room to room. Each time they would enter a room the demons would flee to a hiding place. One by one dad would call them out and the angel would hold them.

When dad and his angel needed more help another angel came, dad told me it was my angel. Dad was very surprised when he called a fierce tall looking demon out of the woodstove. It was much taller than a human being, but it was subject to him in Jesus' name.

When dad came to my room, he was surprised the room was very bright and there were no demons present. He also noticed the presence of Jesus was hanging on the wall like a picture.

When dad told me this, I remembered one night when I actually slept in my own room and in my own bed. I had been so scared that night so I prayed that God would fill my room with light and that He would send angels to remove every demon and that His presence would be with me in the room.

I was so scared that night I was praying anything I could think of to pray. I felt absolutely no different after the prayer and didn't even realize that my prayer had been heard let alone answered. I was astonished to find out that the prayer was answered though because dad saw it in the spirit realm.

So many times, I have prayed and felt nothing and thought my prayers had not been heard or answered. I realized when dad told me about my room that our God listens to our prayers, and He is more than willing to send us help when we need it.

Dad also noticed darkness coming from the speakers of the record player. This made me wonder because I only listened to Christian records. A few nights after dad told me this, I had a dream that revealed the reason. In the dream I was in my room worshipping God with my records. Then a new record dropped, I recognized the song being played. In the dream when this song played, I suddenly felt fear. When I woke up, I knew why the darkness was coming from the record player and which song was the offensive song. Although it was a Christian album this particular song was not about the Lord. I think it was about the singer before he was saved. I threw that record away.

After dad and the angels finished cleaning the house of demons, they took them out of the house. Then the angel with the sword placed an angel at each door of the house to protect it. He also instructed dad to be careful what he let in the house. This made a huge difference.

Then God and my dad tried to talk to me about fear. I had a dream. I dreamt I was in my bedroom on the second floor of our house looking out the windows. There were huge black dogs outside barking and looking up at me. I was definitely their target. My dad was standing outside in front of the house between the house and the dogs. He was not afraid of the dogs. Even though I was upstairs on the second floor I was still terrified. {Ever since I was a little girl, I had always been afraid of dogs.} Then the scene in the dream changed. My dad and I were talking, and he pointed out a little blemish on my face. He said, "It appears to be a little thing." Then he pulled it out of my face it was a huge thing with huge roots that went very deep "But it is a big thing," he explained in the dream. I woke up.

The next day dad asked to talk to me. He talked to me about fear. He said it may not seem like it is a big deal, but it is a big deal and I needed to overcome it. I thought of the blemish in the dream I knew then the blemish was fear, and the roots were deep. The dream had prepared me to really listen to what my dad was saying to me.

The next day at church the sermon was on fear. I not only had fears of the devil, but every other kind of fear also. I had fears on every level of my life. The roots of those were the devil also. The devil is a bully and a bully's power is to scare people. The devil was bullying me through fear. Fear controlled my life. I had many more battles to fight in the future, before I would overcome the fear, this time without my dad.

CHAPTER FOUR

Struck Down but Not Destroyed

We are hard pressed on every side, yet not crushed; we are perplexed, but not in despair; persecuted, but not forsaken; Struck down, but not destroyed.
2 Corinthians 4:8-9

I want to skip ahead to when the big battles take place. We left off when I was sixteen; I want to move ahead to age twenty-three so let's catch up.

I was married when I was nineteen, to my husband, Jim. Mom and dad divorced shortly after, and my sister Carol also got married. In this part of my life, I have a little boy who is three named Jamie and a baby girl named Lonna. The four of us live in an apartment in a very old building, over an insurance company on the main street of a tiny town.

This demonic battle started when I was fourteen.

Through the years of this I became super sensitive to the spiritual realm on the dark side. I had seen many demons but at this point in my life I had never seen an angel. {Although I did have an experience where I heard the angels all night, one night.} If there were a demon attached to an object, I could sense it, or if one came in the room, I could sense it. I became overly sensitive to evil.

Like one time when I first was married, my husband brought home his brother's hat. It kind of looked like Crocodile Dundee's but not quite. The hat had almost a lusty feel when you looked at it, a darkness. The moment I saw it I told my husband I didn't want it in our house. I asked him to get rid of it, but he didn't want to because it was his brother's and he only borrowed it.

An evil spirit would come out of that hat and torment me at night. It looked like a skeleton with long hair. Even though it made my husband and his brother angry with me, I had to throw that hat away.

Certain toys and books would even have demons attached to them. It is like they would have eyes that looked at me. I would tell people from church that their kids' toys had demons on them. {This did not go over well.} Certain conversations would attract demons. I would feel them come in the room; it would make the back of my neck burn. Sometimes I could vaguely see them in the room.

I didn't like this. I would complain to the Lord. "I don't want to see demons; I want to see angels."

I believe this super sensitivity was part of the oppression. At this time in my life, I watched little or no television {I still rarely watch it}. I spent my days listening to Christian teaching tapes, worship music, teaching my kids the Bible, reading Christian books and my Bible. I tried to consume myself with the Lord and nothing worldly. I was sick of evil!

So now, in my story, I am twenty-three and I am grown up {sort of} and I am married with children. During this time, I was pressed in on every side.

First of all, my marriage was a spiritual battle, {a fierce spiritual battle which I wrote about in my first book *The Impossible Marriage*}. In my marriage I felt overwhelmed and seemed defeated, but I wasn't. I was pushing Satan back one difficult step at a time.

Secondly my own mental health was shaky. I was having issues with anger toward my little boy. I was depressed and slipping. I was wearing the same clothes every day. I had stopped wearing make-up, I wasn't even brushing my hair.

We were in extreme poverty and had very little groceries. I was having many battles, battles to keep my husband out of prison, battles for his life, and of course battles with demonic oppression.

I want this book to be about more than just battles about demonic oppression. Satan is the author of every evil such as poverty, sickness and depression, confusion, and temptation. I am starting with oppression because it was a big battle in my life with the enemy.

So, at this time in my life I was pressed in on every side, the inside, the outside, every part of my life. I was beset with fears and problems.

The oppression at this time in my life was especially heavy. I would be fine during the day taking care of my kids and spending time with the Lord, but as soon as it would start to get dark, I would start feeling dread coming on.

As I would lie on the bed at night, I would see big spiders on the walls. They were about eight inches long and were made out of shadows. I hated seeing those spiders; I would pray and rebuke them, but they just ignored me.

Sleep was pure torment. My mind would wake up, but my body would be trapped asleep. I would feel paralyzed and fight to wake up. Some nights I could pray it off and get a little sleep and others I couldn't.

Every night was a battle. I didn't feel like the Lord had left me, He gave me some wonderful miracles in other areas of my life, {I tell more in another chapter}. He would fellowship with me during the day, but night times kept getting worse.

One way the Lord helped me during this time was potty training my little boy, Jamie. I kept the kids in bed with me because I was afraid to leave them by themselves at night and I was afraid too. I asked the Lord to wake me up if my son was going to wet the bed. The Lord woke me every single time. He would gently nudge me awake to take Jamie to the potty. If I would get lazy and not get up, I would get wet.

It was also during this time that my sister, Carol, prophesied to me in that powerful way that I wrote about in the introduction. His booming voice told me that Satan had no power over me.

The oppression continued to get worse, one night the wall was literally covered with spiders. I couldn't get them to leave, so I called my husband in from the living room for help. He was frustrated, "There are no spiders on this wall!" he hollered banging on the wall. The spiders were jumping like crazy to miss his fists as he pounded on the wall.

If I would try getting prayer at church, it seemed to make it worse instead of better. I felt like I was getting punished for getting prayer. I stopped asking for prayer about it.

I remember one day I was reading a book of prophecies; they were like messages written from the Lord through this author. One of them mentioned a delivering angel.

I stopped reading and asked, "Lord, You have a delivering angel!? Send him to me. Send him to me tonight I want him!" All that day I kept telling the kids, "An angel is coming tonight."

He did. He came that night. I felt a presence next to me that felt like pure electricity. I didn't see him, but I felt him standing there all night, it felt like a million jiggawatts of energy. I felt the darkness struggle with him and leave. I didn't have trouble that night. But he didn't come every night; I still had trouble.

Sometimes I would wake up in the night with this horrible urge to hurt my children, it was awful. It scared me so. It felt like I was going to be overpowered and hurt them.

I had read in a book by Kenneth Hagin {I think I have read everything he has ever written}. I read if the devil tells you to do something do the opposite. So even though it felt like I was going to lose control of my hands and hurt them, as they were sleeping, I would take my shaky hand and stroke them gently and whisper over and over "I love you."

It worked, doing the opposite worked. That only happened a few times and then it didn't happen again.

But the oppression got worse, weirder things kept happening and more frequently. I had stopped telling anyone, and I wouldn't think about it during the daytime.

One night I was sound asleep and suddenly I was wide awake with my eyes open. There was a dark being hovered over me with huge red eyes boring on my face. I could tell he looked shocked when he realized I was suddenly awake and disappeared instantly.

Another night I was lying in bed praying, I was praying blessing over my family. As I prayed, I heard the voice of someone I hadn't seen in many years, I knew him before I got saved and I knew he was into witchcraft. I heard his voice, in my room, cursing me as I was praying blessing.

I wondered if this had anything to do with what I was going through. It continued to get worse. I started to see scorpions in the bed made out of the shadows and I would literally feel them stinging me! I would feel the stinging and then see them on my pillow. It was so horrible I didn't even want to try to sleep.

THAT WAS IT! I HAD ENOUGH! I decided I was

going to go to God and tell Him that I had had enough. I know the devil has to go in Jesus' name, so why wasn't he going? I decided I was going to the throne of God and take care of this once and for all.

The Bible says we can come boldly before Him with confidence {Ephesians 3:12}. I started marching. I closed my eyes in prayer determined to march straight up to the throne. I saw myself marching in a grey cloud of smoke. That was all I could see was the smoke, but I kept heading forward declaring I had the right to come before the throne of God. I couldn't find it, but I wasn't giving up. As I was marching, marching through this grey cloud, all of a sudden, I bumped into someone.

It was Jesus! I was so glad to see Him. I wanted to tell Him all my problems. I started to tell Him what was going on but before I got very far, He started running. He ran before the Father and He threw Himself face down before Him and began fervently interceding for me.

I was amazed. I fell into a beautiful peaceful sleep for the first time in a long time and slept through the night. I thought it would be over, but it wasn't yet, but the answer was on its way.

Chapter Five

The Christian Exorcist

Then the seventy returned with joy, saying, "Lord even the demons are subject to us in Your name." And he said to them, "I saw Satan fall like lightening from heaven. Behold I give you the authority to trample on serpents and scorpions and over all the power of the enemy and nothing shall by any means hurt you." Luke 10: 17- 19

Our church had a library a couple doors down from our apartment we were living in at the time. I love to read Christian books: I was reading my way through the library.

A couple days after Jesus had interceded for me, I felt drawn by the Lord to a particular book. I usually stayed out of the spiritual warfare section. I didn't want to think about demons during the day, but this time I felt the Lord point out a book for me to read from that section. I brought it home and read it. It was by a man, I'll call Mr. B, who called himself a Christian exorcist.

The book was a very radical book on spiritual warfare. It had Mr. B's address in the back, so I wrote him a letter and I asked him for help. I was surprised about a week later to get a reply from him. He wrote to me to call him immediately. He wanted to help me.

We made an appointment to drive and see him the next weekend. He was only a few hours away. We were so broke at this time; this wonderful man even provided a babysitter for us and all we had to do was come.

The couple of days before we went things really got crazy. I was even seeing things in full color not just shadows. The night before we left it was dark outside and the wind was blowing as I was looking out the window the leaves on the tree formed two lions that were looking into my window. Then they turned back into leaves. I couldn't wait to get help!

We arrived at Mr. B's that Saturday. The deliverance service was the most fascinating thing I ever went through. There were about five of us going through deliverance that day, my husband, Jim and I included.

First Mr. B led us through a prayer renouncing any involvement with the occult or any involvement with abortion. After that we took communion then he prayed a general deliverance prayer.

He told us if we could see in the spirit, we would look like chimneys with stuff coming out of our heads. As he prayed, I felt something coming up and out. I felt for a few minutes like I was going to start screaming but thank goodness I didn't. The evil I felt passed up, then out.

After that he told us that some evil spirits had to be called out by name. To assist him he had two ladies helping him. They were praying and hearing from the Lord what was going on in the spiritual realm. They would tell Mr. B what was in each person and what needed to be done.

For several hours he worked on us. I could feel everything happening as they prayed. They addressed the thing on the back of my neck that would burn when a demon was present; they saw it and got rid of it. I never felt that burning sensation again.

Another thing that stood out was when they were praying for our marriage. They put a circle of light around us. I felt light zip around us as they prayed. It was an amazing experience. I felt such power as they prayed. The prayer time lasted for hours, but it was over too soon.

All the way home I worried, what if this was my one chance to get free and they missed something? I needn't have worried because there is no distance in prayer.

When we got home, demons were manifesting. We weren't through yet. I kept in contact with Mr. B over the phone. I was in a spiritual fury for several weeks.

My husband was going through huge changes also. I had noticed that sometimes when he got angry his face would actually change. He would go into a rage that would terrify me. His face would take on a caveman appearance, when he would go into these rages, one time his appearance changed so much I didn't even recognize him. After we got home this rage started manifesting in him. I called Mr. B about it and he and his team kept praying for us. Mr. B told me that he and his prayer team cast six savage beasts out of Jim. I never saw his face change again. I had never heard of such a thing, but I was grateful that Mr. B knew so much about deliverance.

Weeks went by as Mr. B and his team worked on me. We kept in contact by phone. I could feel constant warfare around me.

I asked Mr. B why I couldn't cast out those spiders I knew they had to go in Jesus' name. He told me they weren't evil spirits; they were hallucinations.

He also told me he had never worked on someone who was under such a heavy spiritual attack as I was. I was his worst case, ever. I believe it was a few things causing my problems. One was a generational curse that came from my grandma, also there were witches from my past cursing me and thirdly was the fact I was so sold out to darkness before I got saved. Satan hates to lose a soul he thinks he has.

After several weeks while I was lying on my bed, my children were sleeping next to me. It was in the afternoon, it was our nap time, when walking across the wall came the ugliest spider I had seen yet. Its body was at least four feet across, it had very short, tiny legs and it was covered with warts and ugly things. I was so tired of this stuff, and I wasn't sure if it was a demon or a hallucination, so I just rolled over and was going to ignore it.

The Lord spoke to me. "Talk to it." He commanded.

"I don't want to talk to it." I answered.

"Talk to it" He restated so firmly I didn't dare disobey.

I opened my mouth to speak to this spider, but the Lord's words came out instead. He told this thing, that I belonged to Him and that all of His resources were behind me. Then He told it that everything He had and all of heaven and every angel were behind me, and nothing would be held back. He told it that it could no longer attack me and if it resisted Him again, He was going to put it in the lake of fire before its time.

I spoke to the thing, but it was God's words coming out of me. The thing waddled away, and everything stopped. The end had finally come. I had no more trouble, no more spiders or night terrors. I was finally free.

I still had trouble being afraid of the dark and bedtime, for several more years. During a time of prayer and fasting God delivered me of the fear of darkness and going to bed at night. I love nighttime now. I feel God's presence when I go to bed at night.

I belong to Jesus, I belong to Jesus, I belong to Jesus. Do you know what that means?

It means the power of the universe is behind me. I have nothing to fear. He backs me up with everything He has.

I am also thankful for Christians like Mr. B. He helped me at his own expense and never asked me for a penny. I am so grateful.

This particular battle with the enemy was over. I

still had different battles ahead, but this one was done. Speaking to this evil spirit with the Lord's authority finally got me free and much help from a brother in Christ, Mr. B.

Every Christian needs to know something about their adversary, the devil, and how to defeat him and every Christian needs to know their authority in Christ.

CHAPTER SIX

Our Authority in Christ Jesus

"And these signs will follow those who believe: In my name they will cast out demons; they will speak with new tongues; they will take up serpents; and if they drink anything deadly, it will by no means hurt them; they will lay hands on the sick and they will recover." Mark 15: 17-18

And they overcame him by the blood of the Lamb and the word of their testimony, and they did not love their lives to the death   Revelation 12:11

Jesus called Satan, the god of this world. *{John14:30 and John 16:11}* Satan has his kingdom of darkness and evil set up on this planet. Jesus called it the world, or the kingdom of the world, and it has a whole set of rules and values that are opposite of the kingdom of

Heaven.

Those who do not belong to the kingdom of God, but to the world, are blinded from the truth by the prince of the power of the air. The Bible tells us about this *And you he made alive, who were dead in trespasses and sins, in which you once walked, according to the course of this world, according to the prince of the power of the air, the spirit who now works in the sons of disobedience, among whom also we all conducted ourselves in the lusts of the flesh, fulfilling the desires of the flesh and of the mind, and were by nature children of wrath, just as the others. Ephesians 2:1-3 And But even if the gospel is veiled it is veiled to those who are perishing, whose minds the god of this age has blinded, who do not believe, lest the light of the gospel of the glory of Christ, who is the image of God, should shine on them. 2 Corinthians4:3-4*

We, who know Christ, were once part of this kingdom of darkness. We were blinded from the truth of the gospel, by the prince of the power of the air, but now we are part of the kingdom of light.

For you were once darkness, but now you are light in the Lord. Walk as children of light, Ephesians 5:8

The Difference Between Kingdoms

I remember when I came to the Lord. My mind was no longer blinded by the prince of the world. I changed completely. The change was unbelievable. It was like waking up from an evil spell.

First of all, instead of feeling constantly empty and depressed, I felt peace. God was real, alive and I could feel Him with me. I could hear His voice and feel His love.

Before I was saved the name of Jesus caused a bad reaction in me. I didn't like to hear it. Now the name of Jesus was music to my ears. And speaking of music, I could no longer stand the stuff I used to listen to; I threw all my records in the trash {where they belonged}. I replaced them with Christian music.

And I had always been a litter bug, after I got saved, I spit out a piece of gum onto the road as I was walking home from school. I couldn't leave it there, it felt like a weight pulling my heart. I had to go back and pick it up. It had never bothered me to litter before.

I was so different; my mind was literally unblinded and I was seeing things for the first time. Even colors, I had never noticed how beautiful flowers and trees were. The things I used to think were cool, I now hated.

The kingdom of light was a wonderful place to live. I now hated the things of darkness, where before I lived in darkness, and I was progressing continually further into darkness. The light had repulsed me. Now the darkness repulsed me, and I loved the things of God more than anything. Jesus did this for me; He paid a price for me to be set free from darkness.

Jesus Death and Resurrection is Our Key

When Jesus came into this world, in a legal way

through birth, He, through His death and resurrection defeated Satan. He overcame him, and He overcame the world. *Having disarmed principalities and powers, He made a public spectacle of them, triumphing over them in it. Colossians 2:15 "These things I have spoken to you, that in Me you may have peace. In the world you will have tribulation; but be of good cheer, I have overcome the world." John 16:33*

When we receive our salvation through faith in Jesus, because of His death and resurrection, something very important happens to us, we change kingdoms. *He has delivered us from the power of darkness and translated us into the kingdom of the Son of His love. Colossians 1:13*

Translated means immediately zapped, like the teleports on Star Trek. You are no longer a part of the kingdom of darkness you are for now and forever a part of the kingdom of light. You are a citizen of Heaven now; you belong to Jesus and Satan no longer has a claim on you. But he has become your enemy, he will try, but his authority over you has now ended. The key is Jesus. He has legally redeemed you. Your part is faith in Him.

This is all through Jesus and what He has done for us. We are no match for the enemy; it is only in Jesus we have hope. Jesus has left us everything we need through what He did for us on the cross. He left us authority over evil spirits on earth, while He takes care of the evil spirits in heavenly places.

Here on earth, we are behind enemy lines, and we are God's ambassadors here. In our lives and in our families and in those the Lord guides us and prepares us to help; we can take authority over the enemy.I had quite a struggle for many years with oppression. It certainly wasn't an easy battle, but eventually that battle was won. We need to be aware we have an invisible enemy and be aware of his tactics so we can resist temptation and rebuke him, when need be, and to overcome him in our lives.

Be Prepared Before You Fight

I want to explain a couple of things here. First of all, we do not go and start a bunch of spiritual battles. This is a real war, and we need to be led by our General the Holy Spirit. I believe people get a little teaching on the devil and get into trouble. There are some demonic spirits that are tough to beat. We don't just start wars. We don't just start attacking anywhere.

In her book *Tormented Eight Years and Back*, by Peggy Joyce Ruth, the author, Peggy, describes how she naively got into a spiritual battle and suffered for many years with demonic oppression because of it.

Peggy, a young wife and mother decided to share her faith with a Buddhist woman who moved into an apartment near them. Peggy went to speak to the woman out of a sense of Christian duty.

The woman sat in her dark apartment burning incense as Peggy arrived to talk to her about Christianity. Peggy said she knew foundational truths about salvation, but she knew nothing about spiritual warfare. The woman agreed to listen but told her then she would get equal time to talk to her about Buddhism.

Peggy, in a few minutes, told her about Christianity. Now it was the woman's turn.

As the woman talked, darkness attacked Peggy's mind and Peggy's mind began reeling. She became confused and disoriented. She didn't realize she had taken on a spiritual battle, and she was unprepared for the demonic attack she came under.

From that day on, she suffered terribly from demonic oppression. Her mind became oppressed. Her mental condition became so bad she could not be left alone and eventually she was even given electric shock treatments.

The doctors offered her no hope. What she needed was spiritual help!

She continued to suffer until she and her husband became involved with Christians that knew about spiritual warfare. They helped her get free. Now she is completely well and teaching others about spiritual warfare. Her book offers others hope, but before she got free, she suffered for many years.

Yes, we have authority over evil spirits, but we are not to go out on our own, like Peggy did, and start battles we cannot handle. I have seen couples go out to the mission field without taking into consideration the spiritual battle they would be facing. I have seen sickness and divorces happen to people in mission work, who were attacked by the evil spirits that ruled the areas they went into.

That is not to say people should not go into missions, but they need to be led by the Lord and be covered with prayer and be prepared for spiritual warfare. Jesus took His battle with the enemy seriously; He fasted forty days before His showdown with Satan in the wilderness. Fasting is a powerful weapon in spiritual warfare.

Stand Your Ground for What is Yours

Having said that, it is a completely different thing when you or a member of your family are already being attacked by the enemy or if you are being held captive by the enemy in some area of your life. You have no choice but to fight. You are already involved in a battle and now you fight. Don't allow it, don't quit, be bold and use your authority in Christ. You can go forward confidently in the authority Jesus has given you and with the weapons He has provided *{Ephesians 6:10-18}*, knowing that all of heaven is behind you.

Also, God will lead you to help others and fight for others when He knows you are ready. Like Mr. B did for me. Be sure to be led by the Lord.

No one wants a struggle with the enemy. I sure didn't, and I didn't win it overnight. It took a long time. I didn't ask for it, but I got stronger and stronger through it. Jesus gave us His authority on earth because He knew we would need it. We are not alone. We have all the power of Heaven behind us, because of Jesus.

CHAPTER SEVEN
Don't Give the Devil Permission

"And these signs will follow those who believe; In my name they will cast out demons; they will speak with new tongues; they will take up serpents; and if they drink anything deadly, it will by no means hurt them; they will lay hands on the sick, and they will recover." Mark 16: 17-18

God taught me a principle called, "Don't give the devil permission." He taught me his principle, about not giving the devil permission, when I had only been married about two years, and my first child Jamie was only about six months old.

I was reading a Christian magazine and there was a question-and-answer section where readers write in their questions. One lady wrote in a question about healing.

She asked "I believe in divine healing, and I always pray to be healed, but I got ill suddenly and needed emergency surgery. I had no time to pray or believe God, it all happened to fast. What about healing in that situation?"

I don't remember the answer the magazine gave her, but it didn't satisfy me, so I asked God.

"What about that God?" I waited for an answer, but I didn't hear anything, so I forgot about it until later because this time the Lord brought it up.

My little boy was crawling, he crawled up to the electric socket and was about to stick his little finger in it. I yelled "No no!" I gave him a little swat then added "That could kill you."

The Lord spoke, "Right there, that's your answer, Summer. Satan has no authority to take Jamie's life even if he does stick his finger in a light switch socket, but you just gave him permission to, you gave him permission with your words."

I realized that Satan is after our words. Our words carry power and authority and just then I gave the devil permission to take my son's life if he had an accident. I quickly repented and took back those words. I tried to clean up my mouth. I didn't want Satan to have any permission to hurt my children. If I saw them doing something dangerous, I would say.

"No no, other people have gotten hurt doing that,"

My next child came along, Lonna, a beautiful baby girl. When she was little, she put everything in her mouth. It was awful. I was careful not to give the devil any permission over her either.

One day when we were in the church library, she got into a desk drawer and got hold of a bottle of white out. The library worker quickly snatched it from her and said, "No no, honey, if you eat that it will kill you, that's poison."

Her words hung heavy in the air; I felt the Holy Spirit was grieved. I picked up Lonna and headed for the bathroom. I took her in there and prayed. "I command those words to fall to the ground. Lonna belongs to Jesus; she is protected by the Lord. She will not be harmed if she drinks any poison." It had just kind of gotten to be a habit since the Lord taught me not to give the devil permission with my words.

The Principle Put to the Test

Testing came. Jim and I had decided to take the kids to the circus in Flint, Michigan. As we were pulling into the circus, I heard Lonna give out a little cry. I turned and looked, she had a bottle of dry gas in her hand, and she took a big gulp, she cried out as she swallowed it because it tasted so horrible.

My husband had winterized the car the week before and a bottle of dry gas must have rolled underneath the seat. Somehow, she got it open and took a drink. The first thing I did was cuss, and I don't cuss. I was horrified.

"Jim, we have to get to a hospital." I cried.

The only problem was we didn't know our way

around Flint and didn't know how to get to a hospital. I reminded my husband that there was an emergency clinic back in Clio where we lived.

"Let's just go back there so we don't get lost." I suggested. I thought it was an emergency clinic because the sign on the front said Emergency Clinic. So, we turned around and headed back to Clio to the emergency clinic. We got Lonna to the clinic and rushed in and asked for help.

"We are sorry we don't handle emergencies." They told us.

{I wanted to shoot them.}

We were only a couple of miles from home, so we decided to go home and call poison control. Poison control told us to get Lonna to the hospital emergency room immediately and told us how to get there. By this time more than an hour had passed. We hurried to the hospital emergency room but when we got there, they wanted us to take her to pediatrics. We found pediatrics and waited to be seen and they sent us back to the emergency ward.

I want to stop here and say I wasn't feeling any faith at all. I was feeling horror and panic; I was also thinking I was going to sue the Clio clinic and the hospital. My prayer was just "Help!"

We finally got to see the doctor in the emergency ward, and he was very concerned. He told us that particular poison caused blindness. They wanted to start an I.V. to treat Lonna. They didn't usually let parents in, but I was determined to stay with her, and they didn't kick me out because I stayed calm {at least on the outside}.

For over an hour they tried to get a vein to start the I.V. Hours had gone by and the doctor kept saying "We are out of time; it's been too much time."

Finally, they got an I.V. started and treated Lonna. They observed her for a while, while I nursed her to sleep and sent us home.

When I got home, I changed her diaper. When I opened her diaper, I couldn't believe what I saw, there was a little package made out of mucous and in it was the dry gas she had swallowed. The dry gas had never absorbed in her system but had passed through in a package of mucous.

"Lord!" I cried "I had no faith, I even cussed. You healed her and I had no faith whatsoever!"

"You didn't need faith, Summer," the Lord told me, "Satan had no words on Lonna, he had no authority to hurt her."

Watch Your Words

When we belong to Jesus, Satan has no authority to take our lives or hurt us. We have to be careful not to give him the authority. We are living in God's kingdom now; it has different rules. His rules say if we drink anything deadly it will by no means harm us. This sounds easy enough, but it can be difficult. We are trained to say the wrong things.

The Lord taught me more. He taught me when I don't feel good, "Never say it." He told me. "Speaking it gives it more roots. Just say nothing even if you are not healed, if you don't speak it, it will pass quicker, it will have less root."

I tried it, but it was hard because when you don't feel good you want to complain. I started keeping my mouth shut when I felt a cold or sore throat coming on. I noticed it would never completely come on but pass very quickly. This was working great for quite some time then one day I got sick. I had fever chills and felt miserable.

"Lord" I said, "I haven't said anything, and I still feel sick."

"Get up and act like you are well." He told me.

Have you ever noticed in the Bible when Jesus went to heal Jairus' daughter He told the people she wasn't dead, she was sleeping. Was He lying? Of course, not He was speaking against the work of Satan, the author of sickness and death. {Mark 5; 35-43}

Summer Realized Satan Has No Authority

In recent years I work as a home health aide. I was helping a patient I had never helped before into the shower. I thought he could help some, but he was dead weight. When I lifted him I hurt my back terribly. I went home in tears. I thought to myself, "I lifted someone I shouldn't have, and I hurt my back. It was entirely my fault."

A few days later I was doing a midnight shift with another patient. While she was sleeping, I was studying scripture. My back still hurt. I was meditating on *Colossians 1:13. He has delivered us from the power of darkness and translated us to the kingdom of the Son of His love.*

All of a sudden, I saw it, Satan had no right to put my back out. Even though I had lifted someone heavy, I have been delivered out of Satan's kingdom of darkness. He had no right to hurt my back. "Devil you had no right to put my back out." I exclaimed. Immediately the pain left me and never returned.

What good news! We have been translated into the kingdom of the Son of His love! Keep reading, I want to tell you more about keeping the devil out of your life.

CHAPTER EIGHT
Satan Has Territory Stay Off of It

When you come into the land that the Lord your God is giving you, you shall not learn to follow the abominations of those nations. There shall not be found among you anyone who makes his son or daughter pass through the fire, or one who practices witchcraft, or a soothsayer, or one who interprets omens, or a sorcerer, or one who conjures spells, or a medium, or a spiritist, or one who calls up the dead. For all who do these things are an abomination to the Lord, and because of these abominations the Lord your God drives them out before you. You shall be blameless before the Lord your God. Deuteronomy 18:9-13

And have no fellowship with the unfruitful works of darkness, but rather expose them, For it is shameful even to speak of those things that are done by them in secret. But all things that are exposed are made manifest by the light, for whatever makes manifest is light' Therefore He says, *"Awake you who sleep and rise from the dead and Christ will give you light." Ephesians 5:11-14*

People open themselves up to the devil and demonic oppression by messing in the devil's territory. The occult is an absolute no no. Fortune tellers, astrology, witchcraft, horror movies, role playing games, Ouija boards all these and many more are a no no. Some occult things come in cute packages, children's books, video games and toys. Travelers pick up occult items when they travel, little gods from demonic religions. These things cause problems.

One night I woke up and felt such evil in the house my hair on the back of my neck was standing on end. I was trying, at the time, to put my kids to bed to sleep in their own rooms. I started walking through the house praying. I went into the kid's bedroom and my son was moaning in his sleep. There was a toy on the floor that seemed alive, evil was radiating from it. It was so real I didn't want to touch it. I picked it up and threw it out the front door and I told the evil spirit to get out and not to come back. The atmosphere in the house changed. I checked my son again and he was sleeping peacefully. Every once in a while, I have to get rid of something.

Be Careful Where You Go

There are also places, we don't need to go, places of darkness, like bars or casinos. We don't belong to the darkness; we belong to the light. I am not talking about if the Lord leads you into a place of darkness for ministry, which is different. I know Christians have been going into brothels and leading prostitutes to Jesus. They are going into the devil's territory to take spoil. I am not talking about that. I am talking about frequenting places of darkness. I also don't feel that Christians should work in places of darkness.

When my daughter Lonna was a teenager, she started a job at a restaurant I had never been to. She used to visit the gym after work. This particular day at the gym she passed out. Later that night she asked me to pray with her about what had happened that day at the gym. As we prayed the Lord spoke to me. He said, "The devil thinks he can attack Lonna because he says she is working at his place."

I told Lonna what I heard, and she said she felt the same thing. The restaurant she was working at had a bar and was a rough place. She called and quit her job, she also didn't take her paycheck and she picked a charity and asked them to send her pay to them. She handled the situation correctly. Soon, she was blessed with a good job.

Satan Still has Judicial Rights to Present Himself in Heaven's Courtroom

The devil is constantly looking for access into the Christians life. He makes demands from God, when he feels he has rights over you, if you step on his territory. He argues with God and demands his legal rights. Satan is called the Accuser of the brethren.   That means he presents a legal case against you in God's courtroom.

There is evidence of this in the book of Job. Twice Satan goes before the Lord with the intent of testing Job, everything that happened to Job came from Satan.

Now there was a day when the sons of God went to present themselves before the Lord, and Satan also came among them. And the Lord said to Satan, "From where do you come?" So, Satan answered the Lord and said, "From going to and fro on the earth, and from walking back and forth on it." Then the Lord said to Satan, "Have you considered my servant Job, that there is none like him on the earth, a blameless and upright man, one who fears God and shuns evil?" So, Satan answered the Lord and said, "Does Job fear God for nothing? Have You not made a hedge around him, around his household, and around all that he has on every side? You have blessed the works of his hands, and his possessions have increased in the land. But now stretch out Your hand and touch all that he has, and he will surely curse you to Your face!" So, the Lord said to Satan "Behold all that he has is in your power; only do not lay a hand on his person." Then Satan went out from the presence of the Lord. Job 1:6-12

Satan seems to have the right to present himself before the Lord, in a judicial manner, as you can see his purpose is to condemn us. Here we see he, the devil, is behind what happens to Job. He makes demands from God, concerning Job's life, although the Bible tells us Job is innocent. The devil was able to access Job's life by permission. He presented his petition to God. Job was allowed to be tested. The good news is Job does overcome, he stays faithful, and he defeats Satan. In defeating Satan Job took spiritual spoils. He also receives double; of everything Satan took.

In Zechariah 3:1 we see this same thing happening, *Then he showed me Joshua the high priest standing before the angel of the Lord and Satan standing at his right hand to oppose him. And the Lord said to Satan, "The Lord rebuke you Satan!"*

This scene with Joshua is a courtroom and this time Satan lost his case.

If you step on his territory, Satan will take full advantage. He will wave it in God's face that he has legal access to hurt you. And God does things legally. If Satan has gained legal access into your life, God has no choice but to allow him his legal right.

We also see this in *Malachi 3:8-12 "Will a man rob God? Yet you have robbed Me! But you say, 'In what way have we robbed you?' In tithes and offerings. You are cursed with a curse, for you have robbed Me even this whole nation. Bring all the tithes into the storehouse, that there may be food in my house, and prove Me now in this." Says the Lord of hosts, "If I will not open for you the windows of heaven and pour out for you such blessing that there will not be room enough to receive it. And I will rebuke the devourer for your sakes, so that he will not destroy the fruit of your ground, Nor shall the vine fail to bear fruit for you in the field," Says the Lord of hosts; "And all nations shall call you blessed, For you will be a delightful land," Says the Lord of hosts.*

When the Israelites didn't tithe, they came under a curse, the devourer was able to destroy their crops. Satan had power to attack their livelihoods because they were not tithing. God promised if they began to tithe, He would rebuke the devourer for their sake and open the windows of heaven. Tithing keeps our finances in God's hands and out of the devils. Two things change, God stops Satan from stealing from us and God sends us blessings. We want to stay in God's territory where we can be blessed. We don't want any part of the devil's territory.

Have No Part in the Greed of Satan's Kingdom

My sister, also encountered a situation where the Lord told her to quit her job. She was working for a small company that advertised in the back of magazines for people to work at home. People would think they were getting a job at home assembling crafts. They would pay a registration fee and then begin by buying a kit of crafts to assemble. There was really no market for the crafts. The company would only need a few to put in each kit ordered to show how the finished product looked. The company was set up to make a profit off those looking for extra income, but really offered no chance for them to earn extra income.

The Lord woke my sister up in the night and told her His economy doesn't work like this. His economy blesses others. He didn't want her to have any part of this. She obeyed and quit her job.

There is an onslaught in the financial realm of ungodly greed. No longer are people motivated in business by offering others a product or service with the good of the consumer in mind and the pride {not bad pride but good pride} in serving others. This is straight from the evil one. Food companies, to make profits put unhealthy ingredients in the foods that harm people's health. Investment companies have stolen their investor's blind. You see this greed everywhere. CEOs of companies making millions while the average employee in their company can't pay their bills. The new mindset is profit comes first. This is from the kingdom of darkness.

Those living in the kingdom of light have a servant's heart like their King. They seek to offer others a service for the benefit of the consumer because they are motivated by love. The profits they make are blessed by God. To live blessed we need to stay off the devil's territory.

You Can't Successfully Live in Both Kingdoms

I want to tell you about another situation where the Lord taught me about how Satan uses those in his service and how important it is to cut yourself completely out of his service when you come to the Lord.

The story starts when my son brought home a man, he met who had just received the Lord. This man, I'll call him Randy, was in terrible circumstances, he had lost everything including his health.

He had been living in a horrible old camper. The people who owned the camper were abusive to him and he had attempted suicide. My son brought him home and he became part of our family for a while.

Randy was unique, he had been very rich and successful. He had once played guitar for a very famous rock and roll band and had lived it up on the rock and roll scene. He had hob-nobbed with all the rock and roll stars. Randy told me all about life in the rock world he described it as "drugs, sex and rock and roll."

I learned from talking with Randy how Satan operates in his kingdom. While Randy was in service to the devil {playing in a rock band} he was a multi-millionaire, he lived in a mansion with his wife, son and daughter and was chauffeured in a limo and had bodyguards. He had everything anyone could dream of, including fame and fortune.

I want you to see how Satan rewards those in his business. After he no longer played in this band, Randy had one tragedy after another. His son was killed by a drunk driver. His wife and daughter were also killed in a car accident. He lost his parents. His health deteriorated and his money helped along by drug use ran out.

When we met Randy, this once millionaire, barely had clothes to wear. He had been pawning his expensive guitars, and he had very little left. He still had one of his prized possessions, a personalized autographed picture from his friends the Beatles.

I was so glad Randy was a Christian now. I spent

hours telling Randy about all the wonderful things God has done for me in my life, and how He has come through for me again and again.

The only thing was, I wasn't seeing it in Randy's life. He was still in pain he had no money and nowhere to go. He wasn't seeing the miracles in his life that God promises, and I wondered why.

I asked God. This is what God showed me. Rock and roll and its lifestyle are a religion, a false religion. The reason why God's will {which is health, provision, and blessing} wasn't being done in Randy's life was because he hadn't given up his old religion. He accepted the Lord, yes, he had gotten a new God, but he had also kept the old one.

We can't give Satan any territory; we have to get completely out of Satan's kingdom and completely into to God's.

Why?

So, we can live free of him, the devil. This is a goal, yes, but we can do this with God's help. Jesus said, "Satan comes to steal, kill and destroy but I have come that you may have life and have it more abundantly." God's blessings are eternal. Satan destroys those who serve him. Who would you rather serve?

CHAPTER NINE
Don't Be Deceived

So the great dragon was cast out, that serpent of old, called the devil and Satan, who deceives the whole world; Revelation 12:9

Now the serpent was more cunning than any beast of the field which the Lord God had made. And he said to the woman, "Has God indeed said, 'You shall not eat of every tree of the garden'?"

And the woman said to the serpent, "We may eat of the fruit of the trees of the garden; but the fruit of the tree which is in the midst of the garden, God has said 'You shall not eat it, nor shall you touch it lest you die.'"

And the serpent said to the woman, "You will not surely die. For God knows that in the day you eat of it your eyes will be opened and you will be like God, knowing good and evil."

So, when the woman saw that the tree was good for food, that it was pleasant to the eyes, and a tree desirable to make one wise, she took of its fruit and ate.

She also gave to her husband with her, and he ate. Genesis 3:1-6

Eve was deceived by Satan. God made it clear that she or Adam were not to eat of the fruit in the midst of the garden. Satan came along and tricked her; he got her to believe something else. Satan is the deceiver. He works by deception. When God tells us something, we need to stick to it. Eve was tricked by Satan's reasoning. Satan is great at reasoning with us about why we should do something wrong. We have to resist him and to stick to the truth.

Popular Thinking is Deception

Many people believe they are the exception to the rule, God's rules. They think they can live how they please. They have been deceived. Popular thinking does not follow God's laws; it is Satan's deception on society. Murder is called women's rights. Things that God calls abominations are flaunted in the streets, shown on television and even taught as normal to children in schools. In times past popular thinking allowed slavery, which is repulsive to us now. But the new popular thinking is just as bad.

Summer's Lessons on Deception

I have had some real lessons on deception and paid some prices because of it. It happened because I let Satan deceive me after God had spoken to me.

One day early on in my marriage when I only had

one child, my son and he was quite little; I heard a knock on the door of our cozy little mobile home.

As I was walking to answer the door the Lord spoke to me. He said, "Don't let her in the house."

I opened the door just a crack. It was a pretty young girl; she was a Jehovah's Witness. She tried to tell me about her religion.

"I am a Christian" I told her "I don't believe that stuff you're talking about."

That didn't faze her, she kept talking and drawing me into the conversation. She was standing there shivering and shaking; it was a cold wintery day and she looked so young and sweet and cold.

"You are some Christian!" a voice said in my head, {Satan.} "How can you let that sweet young girl stand out in the cold?"

I opened the door to let her in. Out of nowhere came an older much brasher woman who followed her in. Now I had two Jehovah's Witness's standing in my house.

For about twenty minutes the two of them opened their Bible's and argued with me. I am sure they had a different Bible than ours. They were telling me there was no such thing as the Holy Spirit. I was getting angry. Then, when they said Jesus wasn't God, that He was created. I really started getting angry.

To prove it they opened to the salvation scripture *John 3:16 For God so loved the world that He gave His only begotten Son that whosoever believeth in Him should not perish but have everlasting life.*

"See" she said, "He was begotten, He is not

eternal."

I did not have the teaching to know how to debate them but when she took one of the most precious scriptures in the Bible and used it to attack Jesus' divinity, I was filled with rage. I was so angry on the inside; I had the urge to shove them out of my house. They sensed it because they both left in a hurry!

Later that night my little boy, Jamie, and I were home alone when an entourage of demons filed into my house. I felt them and I saw them too. So did Jamie, he didn't talk yet, but he screamed and jumped in my arms. They had encircled the room and were all around us.

I prayed and rebuked and commanded them to leave, nothing happened. They were still there. I felt fear, such fear, so I called my brother-in-law, a Rhema Bible school graduate to pray. They still didn't leave. I didn't know what to do.

I thought of the scripture that talks about whatever things are good and lovely think on these things, so I got out my wedding pictures and Jamie and I looked at them. That calmed us down, but I knew those spirits were still there.

A battle began in my mind. "Is Jesus really God? Where does it say that in scripture?" I was being tormented.

My mother's little Pentecostal church was having a revival that weekend. A special speaker was coming. She was a fireball, anointed Holy Ghost preacher. I went on purpose to get prayer and I did.

When you can't pray through on your own, an

anointed meeting with fired up Christians is a good place to go.

I told the lady preacher what happened, and she prayed. I repented for disobeying the Lord and letting the girl in the house. Under heavy anointing the lady preacher prayed and rebuked the devil. I went home and the demons were gone and so were the doubts in my head.

I had been deceived, just like Eve. I fell for the devil's lie. I wish I could tell you it never happened again, but I can't. About eight or nine years later I fell for a worse trick. It almost cost me my life. I had been a Christian much longer and I should have known better.

We were living in Florida at the time, and now I was the mother of three children, my son James and my two daughters Lonna and Joy. I also watched my two little sisters Maggie and Liz. So altogether I had five little ones.

The Lord had delivered me of cigarette smoking many years before. It all started when I was shopping for someone else, and I had to buy them a pack of cigarettes. As I was paying for the cigarettes, I had a little nudge from the Holy Spirit "Don't put them in your purse."

I started to throw them in the grocery bag, but I didn't want them to get smushed.

"What difference could it possibly make if I put them in my purse or not?" I reasoned with myself. I put them in my purse.

I didn't understand spiritual things. For some reason putting those cigarettes in my purse opened me up to the devil. I didn't realize what was going on, but I started to have thoughts like this.

"I was never overweight when I smoked. Being overweight is just as bad as smoking. I should smoke to lose weight."

At first, I didn't fall for it. Months went by and I was having an urge to smoke. I didn't realize at the time what was going on, but I had opened myself up to temptation when I disobeyed the Holy Spirit and put those cigarettes in my purse.

A couple of months later in a weak moment, I fell for the lie, and I smoked a cigarette and then another. Satan laid a trap for me, and I stepped right in. Suddenly I had opened myself to every kind of temptation. Remember back when I was twelve in my darkness days, Satan's point of entry was cigarettes. Now I opened myself up to everything from my darkness days. It may sound like a little thing, but it wasn't. My whole life had been a war with the devil, who wants to destroy me; I opened myself up for trouble.

I hadn't been smoking for long, maybe a week when I was in my bedroom and I heard a huge crash. It came from the bathroom. I ran to see what happened. The big mirror on our bathroom wall had smashed onto the floor. I stood there in shock trying to figure out how that mirror could have come down. That mirror was deeply anchored into that wall. As I was cleaning up the glass the Lord spoke to me.

"Satan says he is going to take your life."

It angered me. "Well, he is not going to." I told the Lord.

That night I had a severe asthma attack. As I was

going to the emergency room, I promised the Lord I would never smoke again. I received treatment at the emergency room, but I was still having trouble breathing. They sent me home with some pills and an inhaler that I could use every four hours.

At this time my husband and I lived in a duplex, my mom lived on the other side. My husband and mother had jobs. My husband was a waiter at a restaurant. I watched our three kids and my mother's two during the day, so I was home with the five kids.

The next day when my husband and mother went to work, I was a little worried because I was still having trouble breathing. The kids were never as bad as they were that day. I was almost helpless because I was so out of breath.

Things got worse. One of the kids got into some bath powder and got it all over the house. The little breath I had got worse from the powder. I left the kids alone on my side of the duplex and went to my mother's side to get away from the powder.

I was panting so hard I thought I was going to pass out. I sat down on a chair and tried the inhaler again even though it hadn't been four hours. It didn't help. I was sitting in a chair literally fighting for breath. I could no longer get up because it made me too dizzy. I wasn't getting enough air to move. I couldn't even reach a phone to call for help. I just sat there gasping.

As I was sitting there, God sent me help. Two ladies came to the door. I croaked for them to come in. I had never seen them before. They told me why they came.

"We were eating at a restaurant and as we prayed over our food the waiter came up to us and asked us to come pray for you. So here we are."

{My husband Jim was the waiter, thank God he compelled these two ladies to come pray for me, when they came, I believe I was dying.}

They were women of faith. They were Pentecostal like I am. They spoke faith to me and then they prayed over me. I felt no better. I was worried about the kids alone next door and asked them to go check on them. The one lady went and came back. "Oh" she cried "They have the whole house messed up from one end to the other. Then she added "We clean houses for a living we are going to clean your house for you." Then she stopped herself and said, "No I believe the Lord wants you to get up by faith and clean your house yourself, as an act of faith that you're healed."

I caught what she was saying in my spirit and told them I would do it. I thanked the ladies and they left. I stood up by faith and began to walk. I was still panting but I wasn't dizzy. I went next door and hit the powder smell; I was still panting but the dizziness had left.

I checked on the kids. They had settled down and were all five watching television. That in itself was a miracle, God was in control now. I started on one end of the house and started cleaning to the other end of the house. It took me a couple of hours.

I never stopped gasping the whole time, but I still wasn't dizzy. The last thing I did was vacuum up the powder. Finally, I was finished but I was still gasping for air. I decided to try the inhaler again because it had been a couple of hours. This time it worked. The asthma attack subsided.

I had opened myself up to this attack. Satan worked through deception. First when I put the cigarettes in my purse, and I didn't listen to the Holy Spirit. And then when I listened to the devil and started smoking.

For me this was sin. I don't want anyone reading this who hasn't been delivered from cigarettes yet to get into condemnation. My whole life has been a struggle with the devil and for me this was setting myself up for trouble. I fell for his tricks this time. He wanted a foothold in my life because he wants to take my life.

When God speaks to us and tells us not to do something or to do something, we need to obey. We can't let the devil deceive us or we get into trouble. If we learn to obey the voice of the Holy Spirit, He will keep us from Satan's traps of deception.

This is how Satan operates by deception, he got me through reason to disobey the Lord. It almost cost me my life!

CHAPTER TEN

Divine Protection

Because you have made the Lord, who is my refuge, Even the Most High, your habitation, No evil shall befall you, Nor shall any plague come near your dwelling; For he shall give His angels charge over you, To keep you in all your ways. They shall bear you up in their hands, Lest you dash your foot against a stone. Psalm 91: 9-12

You belong to Jesus now and that means you are under His protection. The Lord is completely capable of keeping us from harm and He wants to. I remember when I first got my driver's license at sixteen years old. I was driving like a maniac. I couldn't be bothered with speed limits and things like that; I was in too big of a hurry.

My dad provided me and my twin sister, Carol, with a car to share. It had belonged to his grandfather, our great grandfather. It was a 1963 Ford Galaxy. It had no power steering; the brakes were awful, and the gas pedal had fallen off. {We pressed on the medal thing it had been attached to.}

One day I was zipping around town, too fast as usual, when I almost hit a car. As I was passing the beach I wasn't looking forward. I was looking sideways at a cute boy who was about to cross the street. When I looked forward the car in front of me had stopped. I slammed on the failing brakes and the car came to an abrupt halt. "Good brakes" the cute boy said through the open window.

"That's funny" I said to myself "Usually you have to about drag your feet to get this thing to stop."

The Lord broke into my thoughts. "That is the last time I am going to keep you from an accident." The Lord told me.

"You have kept me from accidents, Lord?"

"Yes, and unless you start obeying the speed limit that was the last time. If you obey the speed limits like you are supposed to, I will spare your car if you are about to get in an accident. No matter what I will spare your life, but if you obey the law, I will not only save your life, but I will also spare your car."

"Okay, Lord" I told Him. "I will keep the traffic laws."

I hadn't realized the Lord had been keeping me out of trouble. I started keeping the speed limits. Driving twenty-five felt so slow I felt like a snail.

God is so good to me; He has miraculously kept me out of accidents like He promised.

One time I was traveling rapidly through an intersection {not over the speed limit}. I was on a state highway and as I came up to the intersection a car ran a red light and entered the intersection from the right. I saw there was no way to avoid impact, the car was right in front of me. I grabbed the steering wheel and stepped on the brakes.

A miracle happened, there was no impact. I am not quite sure what happened but suddenly I was through the intersection and out of harm's way. I couldn't believe I was alive when seconds before I was sure I was done for. I could tell of several close calls but that was the most dramatic. God has kept His end of the bargain and to the best of my ability, I have kept mine.

I did have one incident with a semi- truck but the Lord gave me a dream before it happened so I would know that it was in His care. At the time I was running a paper route seven days a week to pay the bills. My car was vital to our income because without it I couldn't work. I used it to deliver nearly 500 papers a day.

I was worried about my car situation because my car had over two hundred thousand miles on it and it was starting to have problems. I had no savings for a new car or any major repairs and it was on the back of my mind that if something happened to this car we would be in trouble.

I had a dream; I was in a white van and a red van ran into me. "Now what am I going to do" I said. Then I woke up.

Because of the colors, red {blood of Jesus} and

white I felt the meaning was good. Shortly, the dream came to pass. I was sitting in a line of traffic at a red light in my little worn-out car when a big semi- truck going the opposite direction was turning into a parking lot behind me. The back end of his truck didn't clear my car and he slowly started crushing my car on the back end of the driver's side. I beeped my horn, and he realized what happened and stopped.

My car was severely damaged over the back wheel and the tire couldn't turn. The only way I could drive it was with the donut tire on because it was so small. To make a long story short after several weeks of driving on my donut tire, the trucking company sent me a fifteen-hundred-dollar check to cover the damage. I spent five hundred to just get the car drivable and saved the left over one thousand dollars for a new car.

A couple of months later the engine went out on my car and thanks to God, I had some money for another car. God had let me know something was coming and He was in it. He worked it out for good, it was His provision. Of course, I see that in hindsight. While I was going through it, I was wondering why God let my car get hit. Belonging to Jesus is wonderful He takes care of us.

Divine Protection on My Paper Route

When I first started my paper route, I knew it was a dangerous thing to do. We lived in Florida and there was a lot of crime. I would have to pick up my papers in a parking lot at about one thirty or two in the morning. They never came at the same time; you just had to wait for them. I would frequently hear gunshots from the bar down the road. One night there were two drive-by shootings on the same night. One man died, that shooting was a block from my house. The other man barely survived, and that shooting was in the apartment complex where I delivered the papers on foot. Crime was everywhere; one lady paper carrier was murdered.

I was happy to get a job because our finances were desperate. I told the Lord from the beginning, "I have to do this, so I am not going to worry about getting hurt or killed, You take care of me."

I gave it to Jesus, and I refused to worry. I had a little radio with me, and I would listen to Christian radio in the car.

The worst part of my route was a huge apartment complex; it had about forty buildings, two stories high each. I would have to leave my car running, pull up and deliver a few papers to each building and then pull up and go to another building. It took twenty or thirty minutes to do, depending on how big the papers were.

One time while I was delivering in there, police helicopters were circling they were looking for an armed robber. They found him in there where I was, in the parking lot of the apartment complex.

I was protected; the Lord showed me He had assigned a large angel to go with me. My angel was sitting on the roof of my car, and he went with me every night. When I had the Nissan Sentra, he took up the whole roof of the car. I never saw him with my physical eyes I saw him with my spiritual eyes {which is just as good}. I knew exactly what he looked like. He was big, about seven feet tall and he wore a white robe and was very bright. His face shone. If I had to go out on the right side of the car, he would turn his face to the right and watch me as I delivered papers over there. If I had a building on the left, he would turn his face left and I would feel his face shining on me.

The feeling was incredible, his face would minister peace to me, and it was a supernatural wonderful peace. My children could feel it too when they came with me. It was such a wonderful feeling being in that angels watch that I felt very close to God all through my paper route.

This was not true of the first paper route I did though, my first one I hated every minute of every night; I did that one out of obedience to the Lord. I wrote about that in my first book, *The Impossible Marriage*.

As I drove out alone every night, I never was alone and I didn't worry. I knew I was safe. I had my angel with me.

Satan Attacks Summer's Divine Protection

The devil didn't like my divine protection. He came up with a plan to get me out from underneath of it. He used my supervisor, Jill. Jill was my substation supervisor, and she didn't like me. She made it her goal to make my life miserable.

The first thing she did was put the money I collected from my customers, onto the former carrier of my route's account. In other words, she took my money! I was furious and argued with her. I knew she couldn't do that, but she did. I was fuming all through my paper route. I wasn't enjoying the peace like I usually did; I couldn't, I was too angry with Jill.

Every night it was something, she was always doing something to upset me. The Lord showed me my angel again; his hands were bound with a black rope. It was Satan behind Jill; he was using the situation to bind my angel.

Apparently, it was very important that I wasn't filled with anger and upset; it tied my angel's hands. I had to stop and forgive Jill and stop being angry. I also went over her head to her supervisor and got my money back. I had to a few times. But I couldn't go out angry and upset. I had to let it go, I had to, to defeat the devil's plan to spoil my divine protection.

This is a big lesson. I want you to learn this. I want you to see how the devil can get a foothold and mess you up. Money is big with me because I needed it and because

I had to work hard to get it. When Jill messed with my income, she pressed my button. It was not worth it though to lose my protection. Each night I would have to let go of the anger and frustration she caused me {really the devil}. When I saw things from this perspective that Satan was trying to rile me up by using Jill, I let go of the frustration and anger.

After a time, I came in to work one day and there was a new supervisor. I never saw Jill again. I skipped around the substation singing "Ding dong the witch is dead, the witch is dead the wicked witch is dead."

I hadn't realized before how important it is to hang onto our peace and not let Satan steal it. It is so important we will talk about it again in another chapter.

God had another lesson for me to learn. It was this: when we get victory in an area of our lives like the divine protection, I had on my paper route; it is first for ourselves and then it is for others. The Lord told me this when He told me to write my first book, and then He reminded me of what I am about to tell you next. After I received divine protection for me, God used it for someone else. Someone who was going to need help.

First for You and Then for Others

Along the north end of my paper route was a highway. One side of the street had businesses, but the other hadn't developed much yet and was a big field.

In an area, on the field side, someone had opened up a

doughnut shop in a small mobile home like building, or maybe you would just call it a trailer, kind of like what they have at carnivals. The little donut stand would open up at five am.

On Sundays when the papers were especially heavy, I would bring a couple of kids with me to help me. Then for their reward on Sundays, I would stop at this donut stand and buy them donuts. Of course, my husband would want a couple too and I would buy a couple extra to bring home.

My last stop on my paper route before I went home was a beautiful senior complex. It was a high rise, and the apartments were all inside. It had a beautiful lobby with security guards and a fireplace. I would buy the kids their donuts just before we went in there and they would sit in the lobby and eat them and talk to the guards while I delivered papers to the apartments in the building.

This particular Sunday morning my son, Jamie and my little sister, Elizabeth, were my helpers. They were young, only nine and ten. About one am, I shook them awake to get up and dressed and get going; we had a big job ahead of us. I would get ready and out the door in minutes, that way I could sleep longer. I rushed Jamie along, but he couldn't find a shirt. I went in and grabbed a T-shirt from my husband's drawer, it was a large one. It looked like a dress on Jamie, {God was in this.}

My sleepy husband gave me two dollars for the kid's doughnuts and then he gave me another two for his doughnuts and told me which kind he wanted. Doughnuts were a special treat we only did on Sundays. I folded the money and put it in my fanny pack, a special purse I wore around my waist on a belt.

The fanny pack worked well for me on the paper route because I was in and out of the car all night with armloads of papers and I didn't have to worry about grabbing my purse; it stayed on my waist.

We all got going and things went smoothly. When we got to the little doughnut stand it was still dark, around five am. The girl at the stand was young, about nineteen; we would see her every week. I went up to the little window to order the doughnuts but when I opened my fanny pack to order my doughnuts, I could only find two dollars. Where was the money for Jim's donuts? "That's funny" I thought to myself "It has to be here."

I only kept a few things in my fanny pack, my driver's license, keys, asthma inhaler, a few Kleenex, and my money. I took everything out and looked thoroughly; I was determined to get Jim his doughnuts. It just wasn't there. So, I got the kids their doughnuts and headed over to the retirement facility. The kids ate while I worked. After I finished delivering my papers, I opened up my fanny pack to get my keys and there was the two dollars lying right on top.

"Am I going crazy?" I asked myself.

"We are going back to the doughnut stand to get Jim's doughnuts." I told the kids.

It was still quite dark but not completely, the sun was just about to come up. As I pulled up to the doughnut stand the first thing, I noticed was the lights were out.

"What is going on?" I said to the kids, "They can't

be closed."

Then we all saw what was going on. Our doughnut girl was outside with her shirt off. A man was dragging her into the field. When he saw the car, he ran behind the doughnut stand. I stopped pulling forward because I was scared. I thought maybe the man had a gun.

We were all scared. The kids started crying. We wanted to leave but we didn't want to leave the doughnut girl. She had obviously been beaten, her face was swollen, she was half undressed. She had her arms crossed over her chest and she kept doubling over crying.

We were waving for her to come to us and get in the car, but she wouldn't move. She was in shock; we couldn't get her to move. We were all so scared and just wanted to leave but we just couldn't leave without her. She wouldn't come! Suddenly I knew what to do. She was a big girl and Jamie had a large shirt on and she was undressed.

"Jamie!" I cried, "Take off your shirt and hold it out the window." He quickly obeyed. It worked, she saw the shirt and began running toward it. She grabbed the shirt, put it on and jumped in the car. We got out of there as fast as we could.

"You came back." She kept saying over and over "I can't believe you came back." We took her briefly to the police station and then the emergency room at the hospital.

I was absolutely in awe when I realized how God had orchestrated this girl's rescue. He had stopped a rape in progress. First the large shirt and then the missing money; God had all of it planned. He had divinely orchestrated this girl's rescue. Two dollars had disappeared and reappeared. It was no accident my son Jamie could not find a shirt but grabbed his dad's oversized one which perfectly fit this girl.

He showed me it happened because of the divine protection in my life. God for was using it first for me and then for others. It is a divine principle.

Not just the protection but whatever God does in our lives, we bring it to others. We get victory for ourselves and then for others.

God has provided us with divine protection. God gave me divine protection on my very dangerous job. He even saved a young girl about to be raped. You can have divine protection too. It is part of belonging to Jesus!

CHAPTER ELEVEN

The Spiritual Realm

But the natural man does not receive the things of the Spirit of God, for they are foolishness to him, nor can he know them for they are spiritually discerned.
1 Corinthians 2:14

The realm of the Spirit is unseen by natural eyes. Because we don't see it many don't think it is real. Even to those of us who believe, it still can seem like a fairy tale. Remember it is the real world, the one that has always been around.

We live in a plane of existence where we are on a stage being viewed by eternal eyes. Nothing we do or say is in secret. God and the angels see us, the host of heaven called the great cloud of witnesses may see us and we are also being viewed by the world of darkness.

Satan urges us to sin, through demonic temptation. If he succeeds, he relishes in it and brings it before God and accuses you before Him. Then he demands, from God, access into your life because he has legal entry. We can help God out in this battle by being quick to repent and leaving sin alone. Satan so ensnares some, that they can't get out of sin, they are trapped. They need a miracle.

Operate from the Spiritual Realm

We need to keep in mind there is an unseen world all around us that affects what happens in the world we see. When we take our part in the battle between light and darkness, in our sphere of influence, we need to operate on a spiritual level.

Many times, spiritual things do not make sense to our minds. This is why so many don't see the value of speaking in tongues. Our minds don't understand it, but it has a powerful effect on the spirit realm, and it frustrates the enemy. We are releasing the power of God through sound, a very real substance and powerful substance to the realm of the spirit. Tongues are the entry level to the things of the spirit.

Moses won the Battle in the Spirit

Spiritual battles are won spiritually. In Moses day, when he was leading the children of Israel to battle against the Amalekites, Moses was on top of the mountain praying while the people were fighting. The Israelites were only winning the battle while Moses held his hands up. As soon as Moses put his hands down, the Amalekites started winning the battle. Now Moses' arms soon got tired, so

Aaron and Hur found a rock for him to sit on and they held his arms up until the Israelites finished defeating the Amalekites.

This makes no sense to the natural mind! Why would it make any difference whether Moses' hands were up or down? Actually, Moses was fighting the battle in the spirit realm, and it was affecting the natural realm.

Elisha Saw in the Spirit Realm

Elisha was another one in the Bible who understood and saw the spirit realm. In 2 Kings 6:14-16 we read about the whole Syrian army coming after Elisha. The Syrian king wondered why every plan he made the Israelites would know about. His servant told him it was Elisha, the prophet, "Elisha the prophet tells the king of Israel the words you speak in your bedroom."

So, the Syrian king sent his army to Dothan where Elisha was. When Elisha's servant saw the Syrian army surrounding their city he was in despair.

"Alas my master what shall we do?" Elisha's fearful servant asked.

"Do not fear" Elisha answered, "For those who are with us are more than those who are with them."

Then Elisha prayed his servant's eyes would be open, and the young man saw the entire mountain covered with horses and chariots of fire. The spiritual realm was very real to Elisha, and he operated in that realm, a miraculous realm. The whole Syrian army couldn't touch him.

Another time Elisha told the king of Israel to take a hand full of arrows and strike the ground. The king did it three times and stopped. Elisha was angry with the king.

He told him, "You should have done it five or six times." Each time he threw the arrows represented a battle against the Syrians that the Israelites would win.

Does that make sense to our minds, winning a military battle by throwing arrows on the ground?

No! Spiritual things can be very unusual.

A Church Wins a Spiritual Battle

I remember reading a book about an old preacher, his church had been going through hard times. One Sunday morning, up in the pulpit he began to jump up and down and spin around violently. The whole church began to shake, and the congregation was fearful. There was a sense of awe in the air.

What made him jump was, in the spirit, he saw a huge snake up at the pulpit. In the spirit he began to destroy it, jumping and spinning powerfully on top of it. Afterward, the church experienced blessings and growth. He won that spiritual battle against a spirit that had come against the church.

The Spirit World is the Real World

The spirit world is the real world, the one that lasts

forever. It is the invisible world that causes events in our natural world.

By faith we understand that the worlds were framed by the word of God, so that the things which are seen are not made of things which are visible. Hebrews 11:3

Our world came from the invisible realm. We see miracles when we learn, by faith, to reach into the unseen world and pull them into the natural world. Jesus understood this. He was able to heal the sick, raise the dead, walk on water, change the weather and many other miracles.

Jesus has defeated Satan. When we operate in faith in what He has done for us, we operate in victory from a place where Satan is already defeated for now and evermore and where we are already victorious.

That the God of our Lord Jesus Christ, the Father of glory, may give you the spirit of wisdom and revelation in the knowledge of Him, the eyes of your understanding being enlightened; that you may know what is the hope of His calling, what are the riches of His glory of His inheritance of the saints, and what is the exceeding greatness of His power toward us who believe, according to the working of his mighty power which He worked in Christ when He raised Him from the dead and seated Him at His right hand in the heavenly places, Far above all principality and power and might and dominion and every name that is named, not only in this age but also in that which is to come. And He put all things under His feet and gave Him to be head over all things to the church, which is

His body the fullness of Him who fills all in all.
Ephesians 1:17-2
and He raised us up together and made us sit
together in the heavenly places in Christ Jesus.
Ephesians2:6

When we are seated in heavenly places with Christ. Satan is under our feet. But if we come against the devil in the flesh realm, he can defeat us. It is not in our own strength.

Don't Fight from Your Mind

Since I have been writing this book for almost a year now, I have gone through one thing after another. I have had a real battle with the enemy. I had enjoyed divine health for years and would go years without missing a day of work. If I would get sick, I would pray and go to work anyway. God would heal me on the way.

When I started this book, I got kidney stones that wouldn't pass. The pain was unbearable. It took more than six weeks and two surgeries before it was over.

Our finances were depleted by the medical bills. Then our cars kept breaking down. In December we went over one thousand dollars in debt. I kept praying "Lord I want to defeat the devil in this situation."

I continued to feel lousy. I felt like I was dragging myself around. Then my doctor called and informed me I had cancer.

I kept trying to keep these situations in God's hands, but I was trying to figure everything out. How will I pay my bills if I miss work? My mind would work overtime trying to figure out how. I was as worried about money as I was about the cancer.

I called my twin sister, Carol, and told her about the cancer. She hung up and immediately began to believe and pray for my healing. She told me she was doing fine when all of a sudden, her emotions kicked in. She started sobbing and crying and thinking, "I can't live without my sister."

Her husband heard the commotion and came in and firmly asked her, "What are you doing?"

Carol realized in order to combat Satan she couldn't be coming from an emotional realm. She ordered herself, "Emotions get in line with my spirit!"

Immediately the sobbing stopped and she was back to a realm of faith. She encouraged me not to combat Satan with my head but with my spirit. I realized all the conniving and figuring going on in my head had to stop. I was trying to figure out how I would pay for everything if I had to be off work. I was afraid of losing our house.

I committed myself to God and stopped trying to figure things out. I let go.

In every situation I come into, in my life, I want to defeat the enemy and I want God to be glorified. I love to follow my example, Jesus. As it says *in 1Peter2:23 He committed Himself to Him who judges righteously.*

I do not understand why I have been under such an attack, but I am committed to Him who judges righteously. I know when it is all said and done, it will be good.

When Jesus was hanging on the cross, Satan

thought he won, but God had a bigger plan. We can trust Him. Remember when you are in a battle, stay in the realm of faith, where you are seated with Christ in heavenly places. Where Satan is under your feet, and you are coming from a place of victory.

{I am cancer free now.}

Satan s Trophy Room

Spiritual battles are to be fought spiritually. I love reading books from other Christians and learning from them. One such book is *The Priestly Bride,* by Anna Rountree.

In this book she describes an experience where Jesus takes her to Satan's trophy room. There they retrieved a stolen gown that belonged to the body of Christ.

Satan's trophy room is located in the second heaven where Satan has his stronghold. The trophy room, this eerie place that is surrounded by demonic guards is like a museum where Satan displays his stolen treasures. The treasures are stolen from the body of Christ.

Anna sees Miriam's tambourine and other things from the church. The stolen treasures are marked and displayed on black velvet lined tables, taken through the ages and kept in this hideous place. Anna sees some of the displays are empty; that they have since been retrieved by the body of Christ. One such was David's harp.

She also sees, in a gruesome Satanic fresco on the wall, which seems to come to life in the light of flickering black candles. She also sees the empty place where two keys had been, the keys to death and hell. The Bible tells us that Jesus retrieved these from the enemy. {*Revelation 1:18*}

Jesus and Anna passed many evil guards to enter this museum of darkness. Either they were blind to them, or they were temporarily frozen as they quietly passed them and entered.

Jesus leads Anna to a gown on a special display. It is a plain gown that could be worn by a man or a woman. In its fabric she can see it is woven with the attributes of Christ. She can see love woven into the fabric. Jesus explains it is a garment to be worn by the heart. It was present in the early church but had hence been stolen by the enemy.

He motions for her to take it. As she lifts from its stand, sirens and alarms sound and they begin to be pursued by a host of demonic creatures. A chase ensues. Anna jumps on Jesus back and they fly one step ahead of the enraged demonic army.

Of course, they safely make it through the sphere of the second heaven and deliver the gown safely to Heaven.

I was fascinated by her experience. I had no idea there was such a place. I wondered if Satan had stolen anything from me and had it displayed there.

Then one time in prayer, as I was praying with my daughter Joy and my niece Heidi, I entered that place in prayer. I saw the black velvet tables and on them I saw a beautiful crown. I reached out in faith, and I quickly snatched it up. I had no idea what it was for, or what I just did.

Joy and Heidi knew. "It is a crown of victory" they told me. I had been standing in faith for my husband as he fought alcoholism for many years. Although he'd checked himself into many treatment facilities and he never stopped trying, drinking was his normal. {I write about it in *The Impossible Marriage*.}

"Satan has had the victory in your husband's life," the girls continued. "But from this point on you will."

I cheered for joy! It was true. My husband's sobriety increased and increased from that point on. I won a spiritual battle in a spiritual way. When I hear a testimony like Anna's, I know that we can all access the same things. God is no respecter of persons.

An Evil Prison

Another time years before this, I heard a fascinating testimony and then experienced it. This man's testimony was amazing. He told us about how his son would try to live a Christian life, but it wouldn't last. His son had made several attempts to live a Christian life but after a few weeks he would be back to his sinful lifestyle and give up.

This man had decided he had enough; he was going to pray until something changed. He began to fervently pray for his son. As he prayed, he entered a spiritual place. He saw himself enter a prison cell that was in an underground cave. In this prison cell, he saw his son bound in chains and chained to the wall of the cell. There was a demonic guard keeping watch over him.

Boldly in the name of Jesus he demanded the key to his son's chains from the demonic jailer. After receiving the keys, he unchained his son and brought him out of the cell. The vision ended.

He said within two weeks of this experience his son gave his life to the Lord again and has been doing fine ever since. He won a spiritual battle in a spiritual way.

Satan had imprisoned a part of this son's soul in this evil prison. This is where Satan keeps pieces of people souls he has managed to fracture and imprison. This man set his son free in a spiritual way. He entered this spiritual place in prayer and used his authority in Christ. The results were amazing.

If He Can I Can

I thought if he can do that so can I. I entered this place in prayer also. I was looking to see if my husband was imprisoned there, but he was not. I was surprised to see my thirteen-year-old son chained to the prison wall! I had no idea he was going through any kind of problem.

I asked the Lord, "Why is Jamie here?"

"Because he has been listening to music he should not be." The Lord told me.

Later I asked my son about it. At first, he wouldn't tell me and then he admitted it was true. The spiritual realm is real, and we need to remember it affects things on earth.

Another Spiritual Battle Won

I have had many examples of spiritual battles. For many years my husband, Jim, suffered with terrible pain in his knee. He had been in a car accident years before and broken that leg. He walked with a limp, was in constant pain and would moan and groan in his sleep. He even had surgery on that knee, but it didn't help. The pain grew worse. The doctor prescribed Jim narcotics to help him with the pain. This was not good because my husband had a past history of narcotics addiction.

I prayed and prayed for that knee. One day I was praying and suddenly I saw in the spirit what the problem was. I saw an evil spirit in his knee; it looked like a little gremlin. The Lord revealed to me it was a spirit of drug addiction causing the pain.

When the spirit of drug addiction couldn't get Jim to take drugs, he caused the pain and got him on drugs that way. My husband was lying in the bedroom in pain, and I was in the living room. I got so angry I marched into the bedroom and told my husband, "You have to stop taking the pain medicine because it is the spirit of drug addiction causing the pain!" Then I told that spirit, "GET OUT IN THE NAME OF JESUS!!!"

Immediately the pain left him. He was able to get

completely off the pain medication. All the pain medicine in the world wasn't helping him and surgery didn't help him. This was a spiritual battle and had to be won by fighting in a spiritual way.

We need to be aware that to solve spiritual battles we need to come at them in a spiritual way. Not by reason or our own strength. The spirit world is the real world, and it affects the things in our world. When we battle in the spirit we need to come from the place of victory. In Jesus we have victory. He has already defeated the devil. He defeated the devil for us. In Him we are already victorious.

Chapter Twelve

Joy, Peace, Patience and Kindness

Do not sorrow for the joy of the Lord is your strength
Nehemiah 8:10

You will keep in perfect peace, whose mind is
stayed on You, because he trusts in You. Isaiah 26:3

And the peace of God, which surpasses all
understanding, will guard your hearts and mind through
Christ Jesus. Philippians 4:7

But I will hope continually and will praise You yet
more and more.   Psalms 71:14

Satan comes against us in many ways, but he is after certain things. He wants to steal our joy, our peace, our patience, our kindness and our hope. These are treasures of the kingdom of light and that is what he is after. We need to guard our treasures.

He uses attacks against earthly temporal things, but he is trying to take our spiritual treasures. He may launch an attack against you, and you may be in the middle of a storm in your life brought on by the devil, but if he doesn't get your treasures, your joy, your peace, your hope, his mission has failed! You win. You will get back what he has taken plus more. After Job's trouble, the Bible says the Lord blessed his later days more than the beginning.

In the book of Acts in the sixteenth chapter, we learn about Paul and Silas. They were beaten with rods then put in an inner prison and fastened in stocks. They did not lose their joy. The Bible says at midnight they were praying and singing hymns to God.

What happened then? The Bible says there was an earthquake, and the foundation of the prison was shaken; the doors were opened, and the chains were loosed. The keeper of the prison was about to kill himself, but Paul stopped him. "Don't harm yourself we're all here" Paul said.

The jailer took them home and tended their wounds and asked them, "What must I do to be saved?" All in the jailer's household were saved. Satan lost that battle, and he lost many souls that night.

Keep your joy.

Teaching from An Angel

My daughter Joy called me one night. She felt an angel present during her devotion time. She wrote down his words. I am including her entry from her journal that night, in this book. It will help you get free from the enemy and live in God's kingdom. These words touched me.

Joy is……rejoicing before the victory is won, and it presses you toward the victory.

Joy is light in a dark place. Joy can't be held down by gravity {the weight of the world} and doesn't obey the laws of man.

Joy is freedom from present circumstances.

Joy is resilience that never gives up.

Joy is strength that comes unexpectantly to push you on.

Joy doesn't look at the way things are but at the way they will be. Joy confuses and sets fear in the hearts of its enemies. It is something that can't be taken.

Joy appears as insanity but delights itself in the secrets of the Most High, releases power over evil and crushes the heads of its enemies.

Joy links you to the powerful right hand of God.

Joy made Paul run to the chopping block, David dance violently before the Lord. Joy motivated our Savior to endure His suffering.

Joy motivates, illuminates and intoxicates! Joy lifts up, never down. Joy sees things from a higher view, the way God sees them.

Joy knows the hidden delights of the Father. Joy can be felt by the soul but originates from the spirit. Joy is fulfilled in love, joy exercises hope. Joy helps you live, love and carry on. Joy does not equal happiness as love does not equal romance.

Peace is…..the standard that is raised against the enemy. Peace is a line that cannot be crossed, by the enemy.

Peace is more than stillness; it's as waves of the ocean that pushes away worry, fear and doubt.

Peace lets the matter rest in the Fathers hands. Peace continues on in a natural flow of life orchestrated by the creator of all things.

Peace is destructive to the enemy.

Peace immediately takes you from where you are to where you are going.

Peace is a dwelling place, a habitation.

Peace does not depend on circumstances, not on quietness or perfection. Peace comes from trusting God now, not later, from living righteously in the second you are in. Peace embraces the hour, let's go of the past and puts the future in God's hands.

Peace hides you from the enemy. Peace is a goal Satan wants to keep you from. Satan will always have something for you to fix before you can be at peace. Peace says now I will rest my soul in His hands.

Patience is……the fruit of maturity.

Patience is the young woman that wins the man.

Patience is classy, patience is wisdom.

Patience is integrity. Patience seems like the long

way but goes to the top and stays there. Patience does the right thing the right way.

Patience not only says the right thing but with the right tone of voice. It not only listens but listens with a gentle expression on your face.

Patience is trusting God.

Patience gives cooked food not raw meat.

Patience cares about the consequences.

Patience is a rare find in the heart of man, it is bountiful in the heart of God.

Patience rarely gives up.

Patience undoes the damage of rash decisions.

Patience will bring healing.

Patience is constant, it turns faith into fruitfulness. Patience weaves the pieces together.

Kindness is ….the branches in the tree of love. It waits, listens, understands and bestows grace and mercy.

Kindness gives never takes, it leads the way softly, it follows gentleness. Kindness speaks in love, it cares, and it heals. It separates royalty from rags; it's the law of God. He crowns us with loving kindness.

These are spiritual treasures, things of great beauty and value. They are also weapons that defeat our enemies. There are a lot of beautiful things said in that passage but one that stands out to me is that peace immediately takes from where you are to where you are going.

That makes me think of John 6: 18-21. The disciples were crossing the sea without Jesus and the wind came up and they were having difficulty. Then they see Jesus walking toward them on the water and they were afraid.

Jesus tells them, "It is I; do not be afraid." And then the Bible tells us immediately the boat was at the land they were going.

I wonder if they traveled on peace!

We need to begin to realize that the real treasures we have are invisible treasures and Satan will try to steal them from you, things like joy, peace, patience and love. They have more value and power than we can possibly realize at this time.

Remember to guard these things and don't let the devil take them from you. He will try to use temporal things to steal your eternal things, your treasures. Don't let him, it is not worth it.

Chapter Thirteen
Love the Ultimate Weapon

Love bears all things, believes all things, hopes all things, endures all things. Love never fails. 1 Corinthians 13: 7-8

God is love 1 John4:8b

Love is the ultimate weapon. The Bible tells us love never fails. It was love that caused the Father to send His only Son to the world to die for our sins. It was what God did, in love, for us that caused Satan's power over us to be defeated.

There are those Satan has bound so tightly in bondages of hate and bitterness, through those he sends to abuse them, that it almost seems impossible that they could ever be set free. Love is the weapon, the thing that is powerful enough to cut through Satan's chains of bondage and set these prisoners free.

In my own life, Satan thought he had me. It was

love coming from my sister and love expressed to me by the Lord Himself, that set me free.

Love defeats the devil's hold on those that he holds in darkness.

Love Conquers Hate

In the bestselling book *Run Baby Run,* Nicky Cruz tells his powerful story. Nicky was born in Puerto Rico; his parents made their living by practicing the occult. Nicky was given no love and was told he was a child of Satan by his mother.

By eight years old he was so full of anger and bitterness that he had become hardened. He hated his own parents, he despised them. He was violent and the sight of blood made him laugh.

At fifteen, his parents put him on a plane to New York City. There he joined one of the most feared street gangs, called the Mau Mau's and he became their warlord.

Before Nicky turned eighteen, he became a violent street criminal. The Mau Mau's were responsible for many killings. They had declared war on the police and even the police were afraid of them.

All Nicky ever knew in his life was evil, and he was surviving the only way he knew, by HATE.

Into our story comes a young skinny country preacher, named David Wilkerson. It was the late 1950's and the street gangs in New York had taken over. God sent David Wilkerson to New York with a burden for the gangs. He was a naive country preacher, but God was leading him.

David took a stool and a trumpet player and stood up on the street to have a meeting. After the trumpet player drew a crowd, David got on his stool and preached a sermon.

Nicky was in the crowd and so were many other gang members. After the sermon, David called up the leaders of the gangs to shake hands with him.

Reluctantly Nicky went, being egged on by the crowd, he had no choice. Nicky knew how to deal with hate but not with love. Here are his words.

The skinny man walked over to me and stuck out his hand. "Nicky my name is David Wilkerson. I am a preacher from Pennsylvania." I stared at him and said, "Go to ___, preacher."

"You don't like me, Nicky," he said, "but I feel differently about you. I love you. And not only that I have come to tell you about Jesus, who loves you, too."

I felt like a trapped animal about to be caged. Behind me was the crowd. In front of me was the smiling face of the skinny man talking about love.

No one loved me. No one ever had. As I stood there my mind raced back to that time so many years ago when I had heard my mother say, "I don't love you, Nicky." I thought, "If your own mother doesn't love you then no one loves you – or ever will."

The preacher just stood there, smiling, with his hand stuck out. I always prided myself on not being afraid. But I was afraid. Deeply afraid that this man was going to put me in a cage. He was going to take away my

friends. He was going to upset everything and because of this I hated him.

"You come near me preacher and I will kill you," I said shrinking back toward the protection of the crowd. I was afraid and I didn't know how to deal with it.

Even though Nicky threatened to kill David, David pursued Nicky. He refused to give up on him. Later that day David found Nicky again in a basement hang out.

Suddenly there was a commotion at the door, and I looked up and saw the skinny preacher walk in. He seemed so out of place, with his nice suit and white shirt and neat tie, walking into this filthy basement room. He asked one of the boys, "Where's Nicky?" The boy pointed across the room where I was sitting with my head in my hands, cigarette hanging out of my mouth.

Wilkerson walked across the room like the place belonged to him. He had a big smile on his face. He stuck out his hand again and said, "Nicky I just wanted to shake hands with you and . . ."

Before he could finish, I slapped him in the face – hard. He tried to force his grin, but it was obvious I made an impression on him. He held his ground and the fear once again welled up inside me so that I was sick to my stomach. I did the only thing I knew to retaliate. I spit on him.

"Nicky, they spit on Jesus, too, and he prayed, 'Father forgive them, for they know not what they do.'"

I screamed at him cursing, "Get the--- out of here!" and I pushed him backwards toward the door.

"Nicky before I leave, let me tell you just one thing. Jesus loves you."

That night Nicky couldn't sleep. He got up and went outside of his grubby little apartment. The sun was just rising. There was David again. Nicky was amazed when he spoke to him; he seemed to know what was going on inside of him. He told him again Jesus loved him.

David was using the ultimate weapon; he loved this hopelessly lost young man with God's love. He was truly genuine.

Next David did something everyone thought was crazy. He put on a big gospel crusade in New York. He rented a big auditorium and invited all the gangs in the city. He sent buses to pick up the gangs and bring them out of their turf to the arena. Everyone thought the rivaling gangs would kill each other, and they almost did. But God's Spirit was there and something else happened instead. Many were saved and Nicky was one of them!

The first thing the following morning, David received a call from the police station telling him to come down there immediately. When he got to the station, the police were amazed because Nicky and his gang, the Mau Mau's, the same ones who had declared war on the police, had come down to the station to turn in their guns and to ask the police to autograph their Bibles.

The rest of Nicky's story is also incredibly fascinating because Nicky had so much to overcome, and God was so faithful to keep him in His care. Nicky became a powerful minister and still ministers today. Nicky had a long road ahead, but every time Nicky was about to quit,

God would come through in amazing ways.

I heard Nicky Cruz in person many years ago and his testimony is so powerful that people weep as he tells his story, and the altars are crowded with people seeking salvation.

David Wilkerson's passion was the lost young people in New York City. He was sent there by God, and he represented God with real love that was willing to be spat on and struck and threatened. Nothing but God's love through his servant could break through Satan's hold on this young man.

Love is the most powerful force. It can break through chains of satanic bondage to set others free. God commands us over and over again to love. To love our enemies, to turn the other cheek and to bless those who curse us.

I am sure most of us have heard this story, both David's book *The Cross and the Switchblade* and Nicky's book *Run Baby Run* have sold millions of copies. The story was also made into a movie starring Pat Boone and Erik Estrada. This story has been a testimony for many years, to the body of Christ, of the power of God's love through the gospel, delivered in love, to the hopelessly lost. I want to tell you another testimony of God's love, displayed in a powerful way by his servants that rocked the whole world.

The Ultimate Price of Love Conquers Evil

This story also happened in the 1950's. This is the

story of the most feared, and as Time magazine called them, "the worst people on earth", the Auca Indians, and five missionaries that in love laid down their lives.

The Auca Indians were a tribe of people from Ecuador. The most common cause of death among the Auca was murder. They even murdered each other. The Auca's killed for any reason at all. If they even suspected anyone of holding some kind of grudge against them, they would kill them first.

Even the toddlers were taught to kill; they were given spears to play with and would practice stabbing human shape wooden targets. Children as young as six years old would go on raids. When someone was mortally wounded, they would allow the children to finish them off.

No one was safe and all lived in constant fear. Even parents killed their children if they tired of them. If someone became old or sick, they buried them alive. They were a people in deep Satanic darkness, bound in spiritual chains and unreachable. Even their vicious neighbors, the headhunters feared them.

Five missionaries, who had been working with other tribes, wanted to reach the Auca people; they were Jim Elliot, Pete Fleming, Ed McCully, Nate Saint and Roger Youderian.

They got a break when a young girl named Dayuma who escaped from the Auca people, taught them the language. Once the men learned the language, they began to fly over the Auca people and lower presents down to them in a bucket. They sent copper kettles, shirts, buttons and small knives. The people seemed receptive and even began sending gifts back, such as fruit, a feathered head dress, live parrots and even a balsa wood carving of the plane. As the missionaries flew overhead, they spoke to the people with a loudspeaker in their own language, "We like you, we like you. We are your friends."

After a couple of months, the five men decided to land the plane near the Auca camp, on a sand bar in the river. They made camp and were there to bring the gospel to this people. They knew the danger they were facing but all had made the decision to lay down their lives for these people, if need be. Even though they had brought guns they made a decision to never use them against the Auca, even to save their own lives.

And they didn't. All five men were speared to death, by the unstable Auca people.

John 15:12-13 tells us *"This is My commandment, that you love one another as I have loved you. Greater love has no one than this, than to lay down one's life for his friends."*

These five men did exactly that. They loved and laid down their lives for a chance to bring the gospel to the Auca's.

These five men's lives were not wasted. They dealt Satan a mortal blow in his hold against the Auca people; they broke his power with the power of love, the greatest kind of love. The prince of this world was defeated, through five Christians that defeated him just as Christ defeated him on the cross.

To those who don't understand the things of the spirit it may have looked like evil won, it didn't. Love never fails.

The story is not over. Within three years, Elisabeth Elliot the widow of Jim Elliot and Rachael Saint the sister of Nate Saint, had moved into the Auca village and were teaching the people the teachings of Christ. A Christian church was soon established and now about one third of the Auca people are baptized believers.

Something else happened too, these men became a testimony to the body of Christ. The story became known worldwide and many Christians felt called to the mission field through their story. It is estimated that thousands followed these five to the mission field.

There is great power that is released against the enemy through the greatest love of all, as Jesus said, "Greater love has no one than this than to lay down one's life for his friends."

Love is the ultimate weapon against the enemy. Love for those that Satan is holding in darkness. It is the weapon that Jesus used. It is the weapon David Wilkerson used and it is the weapon these five brave missionaries used. Love is the ultimate weapon for all of us to use.

CHAPTER FOURTEEN
Don't Let the Devil Win, Give the Situation to God

Therefore humble yourselves under the mighty hand of God. That He may exalt you in due time, casting all of your care upon Him, for He cares for you. Be sober, be vigilant because your adversary the devil walks about like a roaring lion, seeking whom he may devour. Resist him steadfast in the faith, knowing that the same sufferings are being experienced by your brotherhood in the world. But may the God of all grace, who called us to His eternal glory by Christ Jesus, after you have suffered awhile, perfect, establish, strengthen and settle you.   *1 Peter 5: 6-10*

Whatever the devil brings at you it is possible to come out on top in the end. To accomplish this, it is imperative to keep the situation in God's hands and not to let your heart get full of unforgiveness, bitterness or worry and fear. Although you will be attacked by these things, keep resisting them. If this seems too hard, remember, Jesus left us an example. He did this very thing, so we also can do it, with His help and then we can win the victory in the situation just as Jesus has.

For to this you were called, because Christ also suffered for us, leaving us an example, that you should follow His steps: Who committed no sin, nor was guile found in His mouth, who when He was reviled, did not revile in return; when He suffered, He did not threaten, <u>but committed Himself to Him who judges righteously</u>.

1 Peter2:21-23

At first, I hated that scripture but now I love it. I think it is wonderful, the example Jesus left us. He did not get revenge; He left it to Him who judges righteously. He gave the situation to God!! I want to put out two examples in this chapter, both true stories.

In the first example I will tell you about, God did not get the victory in the situation {but if we learn from it, He will} and in the second one He did get the victory, we can learn from both.

Situation One

The first situation happened in our church, years ago. I just observed this situation and was praying for the people in it, so you're getting this situation from my perspective. I went to a large church, and I loved it. We had a wonderful pastor I will call Pastor Joe. Pastor Joe's son, Alan was in the investment business. Many of Pastor Joe's parishioners invested with Alan, but I will single out one particular of Pastor Joe's parishioners and Alan's investors named Bob.

I met Bob one time when I was in crisis about something and went up for prayer after church. Bob was on the prayer team. Bob was an older retired man who loved the church and loved serving there. He was a kind and compassionate man. I appreciated Bob because of the prayer I got from him. I could see that he was happy and loved belonging to the church.

Many people of the church invested money with Alan and Bob was one of them. Even though he was not well to do, Bob invested his life savings of about one hundred and twenty thousand dollars with Alan.

Bob lost everything, because the investment company was in trouble and was using the money of the investors to pay off earlier investors. Alan claimed to be innocent of wrongdoing and just was following the orders of the owner of the company, a lawyer.

Alan and his boss were arrested and charged with fraud; all together their investors lost about three million dollars.

All the time this was happening I worked delivering newspapers, in fact Pastor Joe was one of my newspaper customers. The whole story was being played out day by day on the front page of our city's newspaper. I felt so bad for everyone involved. In the wee morning hours as I was seeing the papers before anyone and even delivering one to Pastor Joe, I would be praying about the situation, and everyone involved. I would pray for Pastor Joe as I would deliver his paper, he was often on the headline. He was a wonderful pastor, and he was in the middle, between his son he loved, and the parishioners he loved.

I asked the Lord about Alan and all He said was "I have called him to preach." I think he was like Jonah running from the ministry to the business world; of course, his ship would sink. I also believe most of the fault was with the lawyer who owned the company, Alan's boss.

Bob was also in the newspaper articles, he and other investors. I remember seeing a big picture of Bob looking very sad on the front cover of the paper one day. He was angry and bitter.

Pastor Joe and his family were receiving threats from former parishioners, one day as they were entering the courthouse he was almost run over by a car; it was an angry investor. I wondered if it was Bob, he was so angry and bitter he wanted Alan in prison.

We newspaper carriers had to buy our bills from the newspaper to bill our customers for payment. The bills had a small picture of the newspaper on the front of each of them. It just so happened that the picture of the newspaper printed on the cover of the bills was the one with Bob on the cover.

Every time I would write one of my customers a bill, I would see Bob's sad face. I continued to pray for Bob; every time I wrote a bill to a customer. I would see his picture on the bill and say a prayer for him. I understood why he was angry but not so angry that he would become violent against Pastor Joe.

I also continued to pray for Pastor Joe and Alan, every morning around three a.m., as I delivered Pastor Joe his paper, I would pray for them.

The story ended badly for everyone. Alan insisted on his innocence and would not take a plea bargain. He received a very long prison sentence.

I said to the Lord as I saw the paper early that morning "I guess that was your will, Lord."

"That was NOT my will," the Lord said emphatically. "Alan has not yet put it in my hands."

The next bad news was Bob died suddenly. I believe he died of bitterness. "What about Bob, Lord?" I asked. "He lost all his money, it wasn't fair."

The Lord answered me, "If Bob would have forgiven Alan, I would have brought him back every penny he lost."

This situation was orchestrated by Satan to destroy these people. The outcome could have been different. We can't allow Satan to destroy our lives with hate and bitterness. Remember our scripture in 1 Peter, well, let's go up a few verses this time. I want you to really get this scripture because if you do you will defeat the devil like

Jesus did.

For this is commendable, if because of conscious before God one endures grief, suffering wrongfully. For what credit is it if, when you are beaten for your faults, you take it patiently? But when you do good and suffer for it, if you take it patiently, this is commendable before God. For to this you were called, because Christ also suffered for us leaving us an example, that you should follow His steps: Who committed no sin, Nor was guile found in His mouth, who, when He was reviled, did not revile in return; when He suffered, He committed Himself to Him who judges righteously. Peter 2:19-23

How can we suffer terrible loss and have the ability to give the situation to God? We can, if, we see the real enemy is Satan and the way to defeat him is to follow Christ's example.

And I believe sometimes it is just too hard for us to do on our own but if we try God will help us supernaturally. We can't give in to the devil; we can't let him destroy our lives; we have to keep giving the situation to God. We have to defeat him with God's help and by following Christ's example.

Situation Two

In this next true story I will use the real names. This story was broadcast on Focus on the Family, a wonderful radio show that is very positive. This story is of an awful situation that happened to a Christian lady who immediately gave the situation to God and the results were miraculous. This amazing woman in this story did realize Satan was her real enemy and she kept the situation in God's hands. Satan was defeated.

The story is about a lady named Margie Mayfield. She was out Christmas shopping and as she was going to her car, a man came and pointed a gun in her back and grunted "Get in."

He got in the driver's side and pushed her over to the passenger side. He told her to sit on her hands and if she moved, he was going to kill her.

What she didn't know was that she was with Stephen Morin, a killer wanted in over nine states for over thirty cases of rape and murder. Stephen Morin had been on the F.B.I.'s most wanted list for over ten years and had managed to elude the authorities. His last murder had just taken place earlier the same morning; he had killed a young woman that looked a lot like Margie.

The first thing Margie said to this killer as he got in her car was "Do you know Jesus?"

"No, I don't know Jesus!" the killer growled back at her.

Then Margie said, "I am going to pray for you."

Margie began to pray for this mass murderer Stephen Morin; she actually laid her hands on him and boldly began to pray. She laid her hands on him and bound the devil, she told the devil in the name of Jesus that he could not operate through this man, and then with great boldness she declared to the devil that Stephen, her captor, would be serving the Lord by the end of the day.

After she had finished praying for Stephen she went back to sitting on her hands.

"You don't have to do that" he told her, his attitude already changing.

God was in control.

All day as the killer drove with Margie in her car, she told him about the love of Jesus. She felt a supernatural love for him, coming from God. He told her there were roadblocks set up for him all over the city and that he would never go to prison again. He told her he was going to kill himself if he got in a shoot-out with the police.

Margie explained to him if he killed himself, the way he had lived, he would go to hell. Then she told him that he did not have to go to hell because Jesus loved him and took his place on the cross for his sins so he wouldn't have to go to hell.

Stephen told Margie he had never felt so much love from another person in his entire life. He told her how he had been filled with hate and anger since his childhood. Stephen told Margie how his mother had hated him, her own son, and it began Stephen on a life filled with hate.

Margie felt God in control throughout the day and she was not afraid. Margie was even telling Stephen where she thought the Lord was leading them to drive. Stephen obeyed her direction. God was in charge, and He was giving Stephen time to hear the gospel. All through the day they eluded the police.

At one point during the day Margie felt fear. She thought of her husband and children and wondered if she would ever see them again. Stephen noticed the difference in her, and they stopped and prayed again, they put God back into control. After being in the car with this killer all day Margie said that Stephen pulled the car over, put his hands in the air and asked Jesus for His forgiveness.

"It's gone!" he cried, "The hate and anger are gone!"

Stephen was a new creature. Margie continued to teach him scripture and left him at a bus station with a notebook of scriptures to study.

Then Margie drove home and she found the police were waiting at her house. Because she had been missing all day, her husband and the police thought she may have been kidnapped by this killer they were searching for.

She told the police the whole story except where Stephen was. God was still in control, and she didn't feel that she was to tell where Stephen was yet.

Then later that night the Lord told Margie that it was okay to tell the police where Stephen was now. The police found Stephen at the bus station, where Margie had left him. He was diligently studying scripture.

Stephen surrendered peacefully and he told the police if they had found him earlier, he would have put up a fight, but he was changed. Margie could have wound up dead; instead, she acted supernaturally because she gave the situation to God and followed Him. Satan lost and God won.

Stephen Morin was sentenced to death for his many crimes. He served the Lord until his execution and died with a testimony on his lips.

Margie visited Stephen in prison the day before he

died. The warden came to her with tears in his eyes; he told Margie that Stephen had led him to the Lord!

The warden was saved by the death row inmate! The devil just keeps on losing! God won in this situation because Margie gave the situation to God, and she overcame the devil. She was literally staring death in the face, but the situation turned around in a miraculous way.

You can see through these two situations that only when we put God in control, just as Jesus did can we defeat the devil. In the first situation Satan was victorious and there was great loss. In the second story there was a great victory! Don't let the devil win! Give the situation to God!

CHAPTER FIFTEEN
Be Led by the Spirit

For as many as are led by the Spirit of God these are the sons of God. Romans 8:14

Learning to hear and follow the voice of the Holy Spirit is vital to our Christian walk. The Holy Spirit is our Guide. The Holy Spirit is especially our Guide when it comes to dealing with the enemy. He sees the whole picture. If the devil has laid a trap for you, the Holy Spirit will warn you.

You have been delivered out of the kingdom of darkness and into the kingdom of Light. You are no longer under Satan's rule, but you are under rule. God is your ruler. Satan is no longer our master, but he does have a kingdom. We don't go in and take spoils until the Holy Spirit leads us. We are not presumptuous or prideful and go without Him but when God leads us, we may go in boldly.

A Pastor is Led by the Holy Spirit

When we lived in Florida, we lived on the west side of our town, and we lived a couple of blocks from the sex strip. Every other shop was a porn shop or a triple x video shop. Prostitutes also walked up and down the street, stopping cars and getting in.

On the far east side of town was another sex strip, another satanic stronghold and it was worse. There were more porn shops, a park which was made to be a family park, but sex seekers prowled it. It all centered around a bar with nude entertainment. The bar generated so many problems that for years the local police tried to get it shut down. They were not successful. I had been watching the battle in the newspaper for years. Whenever they would try to close it, a court battle would ensue, and the owner of the bar would win.

Into my story comes a very humble man I will call Pastor Rob. He started a small church in a closed down porn shop on the smaller sex strip near our house. The Lord led him there. I went once to the church services and I was hooked. Not only was Pastor Rob led by the Holy Spirit, but he was also teaching his small congregation to be.

There is nothing more exciting than to go to a church where the pastor lets the Holy Spirit lead. They are rare. I would love every minute of those meetings and you never knew how long they would last because Pastor Rob would never stop the meeting while the Holy Spirit was moving.

The workers of the church would go into the back room to pray for the service before it began. Sometimes Pastor Rob would come out of those pre-service prayer meetings so drunk in the Holy Spirit he would have to be helped up to the front and onto the platform.

Pastor Rob would encourage the tiny congregation, "If the Holy Spirit is prompting you to give a word or share something, come up to the front." He would have an open microphone set up for the congregation to use. It was a training time. One young housewife who looked very poor, and she had an extreme southern drawl would come up frequently.

"Excuse me y'all" she would say in her shy southern voice "The Lord's tellin' me to do this. Awmazin grace how sweet the sound that saved a wretch like me." She would sing-say the words to a hymn slightly off key with no music at all. On and on she would go through all the words to the song.

The next week here she would come again; I would inwardly moan. This week it would be a different hymn. "What a Friend we have in Je-esus" again the off-key sing-say through every word she could remember.

I only heard an off-key singer. I didn't see at first the changes in this willing vessel. She was willing to obey the Holy Spirit and sing and willing to grow stronger in obeying the voice of the Lord. Thank goodness Pastor Rob understood. After a while she started coming in with testimonies. The Holy Spirit was leading her through the week also. She was becoming a threat to the kingdom of darkness; she was following the voice of the Lord.

One week she told us "The Holy Spirit woke me up around midnight and told me to go out into the alley. There were two bums fighting out there. I think they were drunk. I yelled 'In the name of Jesus' and the one with the knife ran away but the other one dropped to his knees. That is when I led him to the Lord."

Miracles were happening in this little church. People were being led by the Lord. Another regular attender of the church got up to the microphone. He was a tall, lanky, awkward looking fellow.

"I was at the Seven Eleven" he told us. "I walked up there to use the pay phone outside, while I was dialing the Holy Spirit spoke to me.

'Are you ashamed of the name of Jesus?' He asked me.

'No of course not' I told the Lord.

'Will you shout it?' the Holy Spirit asked me.

'Yes'

"Then go into the store and shout it.' "

{As I was hearing this story, I was glad that it was him and not me.}

"So" this young man continued "I went into the store, stood inside and shouted as loud as I could, JESUS!! I saw a man run out of the store and then the cashier, a young woman started yelling.

'Did you see that! Did you see that! That man was holding up the store. He had a gun to me and when you came in and yelled Jesus! He ran away.'"

That is what it was like at Pastor Rob's church; simple ordinary people were learning to be led by the Spirit.

And then there was Pastor Rob, God could depend on him. Around the corner from the church and my house was a shopping center where I bought groceries.

In Florida it is very common for homeless and transients to hang out in the grocery store parking lots hoping to get a handout from shoppers as they leave the store or to find money shoppers drop.

In our grocery store parking lot, it was no different. I would actually recognize some of the same faces of the transients and bums that seemed to be there all the time. There was a certain destitute looking man that I saw often standing outside the door of my grocery store. I assumed he was homeless.

One day I felt drawn to him. I wasn't quite sure why or what to do so I smiled at him and passed him by.

Thank goodness Pastor Rob was more sensitive to the Holy Spirit than I was. The next Sunday the destitute looking man was in church with Pastor Rob.

The same day that I had felt drawn to this man, Pastor Rob had also felt drawn to him and stopped to talk to him. This man was at the end of his rope and had decided to commit suicide that day. Pastor Rob stopped and talked to him and then brought him home to his house for dinner.

At church on Sunday Pastor Rob told us all about it and he had us all gather around the man to pray. This man felt so unloved that after we prayed for him, the whole church lined up and took turns, one by one we lined up to give him a hug. It was a holy moment. He looked like a wilted flower coming back to life. Instead of suicide this man got Jesus. Thank goodness Pastor Rob had listened to the Holy Spirit.

I noticed something else too: our sex strip was being affected by the spiritual light coming from the little store front church. One by one the porn shops were closing. The darkness was fleeing from the light. Finally, there was not one porn shop or x rated video shop left on our strip. I didn't see the prostitutes anymore either. Light had overcome the darkness but who knew it was because of a humble little preacher, in a little store front church, who knew how to be led by the Holy Spirit.

"The Lord is leading me to hold a revival meeting on the East side of town." Pastor Rob announced one Sunday.

This was big, Pastor Rob rented a big hall because he felt such an importance about these meetings. The Sunday before the big event we all gathered around Pastor Rob to pray. I went up and prayed too.

I had a vision as I prayed. I saw Pastor Rob in heavy armor. He looked like a knight from King Arthur days. His armor almost looked black because it was reflecting off the dark angry clouds that were in the sky above him. It seemed he was preaching to the clouds.

Was it a coincidence that Pastor Rob's meeting was right in the middle of the other sex strip and the hall was near the nude bar?

Pastor Rob was disappointed because a very small

crowd showed up for the meetings and he had rented a big hall.

I wondered…. I knew God was doing something.

I was right. Soon after, in the paper I saw an article, the law enforcement after many years had finally won their case against the nude bar; it had to close down.

Was this a coincidence?

No! God did do something big in Pastor Rob's service. He pulled down the Satanic stronghold that kept that evil place operating!

Sometimes quiet, humble, unnoticeable people pose a real threat to the devil because they are led by the Holy Spirit! Pastor Rob did not set out to close down the very dark areas that the police could not make a dent in, but he did. He was just following the Holy Spirit and teaching others to do the same.

Chapter Sixteen
Forgiveness Beats the Devil

For if you forgive men their trespasses, your heavenly Father will also forgive you. But if you do not forgive men their trespasses, neither will your Father forgive your trespasses. Matthew 6:14-15

Therefore as the elect of God, holy and beloved, put on tender mercies, kindness, humbleness of mind, meekness, long suffering, bearing with one another, and forgiving one another, if anyone has a complaint against another, even as Christ forgave you, so you must also do. Colossians 3 12-13

Forgiveness defeats the devil in our lives. Unforgiveness will only hurt us. It is destructive to our being. Forgiveness frees us from bondage and releases good. It is following the example of Jesus. It is living in His kingdom.

Satan is looking to destroy you; he wants to keep you from all God has for you. He is behind the evil others have done to you. Unthinkable things are done to others. Is forgiveness possible?

Yes, it is, but we sometimes need God's help.

An Amazing Story of Forgiveness

I read a book called *The Lost Child* by Marietta Jaegar. It is an amazing story of forgiveness. Marietta, her husband and their five children, a family from Michigan, went on a camping vacation in Montana.

While on vacation, their youngest child, seven-year-old Susie was kidnapped. The family woke up to a hole in the tent near Susie's sleeping bag and Susie was gone, without a trace.

For five weeks they stayed in Montana while the sheriff and the F.B.I. looked for Susie. They finally returned to their home in Michigan without Susie.

Marietta was filled with rage. Although she didn't know who had taken Susie, she hated him and wanted him to "hang." Not wanting to be consumed by this hatred and because of her faith she made a decision to let go of the hate and forgive the monster that took her baby.

It was not easy. She began to do the unthinkable; she began to pray for her enemy. Many times, the hate and rage would return but Marietta fought it and daily prayed for the kidnapper. Marietta thought of him as a person in trouble that needed help. Over time there was a change in Marietta, as she prayed for the person who took her daughter, the rage subsided and, in its place, she found that she loved her enemy.

A year to the day from the kidnapping the kidnapper called her. He called to taunt her and then hang up, but he was met with true and genuine compassion from Marietta. He was compelled to keep talking with her even though he knew the F.B. I. had the phone tapped.

When she asked the kidnapper what she could do for him, he broke down crying. She kept him on the phone for eighty minutes because of her loving gentle attitude. The F.B.I. was able to put together enough information from the phone call to catch this man.

The kidnapper was David Meirhofer a twenty-four-year-old man who had served in Vietnam and lived in Montana. Marietta's seven-year-old daughter had been kidnapped by him for sex and suffered before the kidnapper killed her.   Still Marietta forgave. Her prayers had changed her heart and she truly cared about her daughter's murderer. Marietta asked the prosecutor not to pursue the death penalty. Only then, when Meirhofer learned this, was he willing to confess. He confessed to four murders including Susie's.

Several hours after David Meirhofer confessed, he committed suicide. Marietta grieved for him because she truly had forgiven him.

Marietta defeated the enemy through her forgiveness. Although nothing could bring her daughter back, she refused to let hate and bitterness destroy her life. What Marietta did was no small thing. She did the impossible, the unthinkable, she forgave something so horrific it is every parent's worst nightmare. Marietta's forgiveness became her life's greatest work and her gift to the world. She blazed a trail of light in a dark world, by living like Christ. Now others can find their way. She dealt Satan a terrible blow.

She also, because of the love and compassion in her voice, which kept the kidnapper on the phone, enabled the F.B.I. to stop a serial killer who had already taken four lives from continuing to take lives.

Many serial killers go on for years cloaked in Satan's unholy protection; they continue destroying many lives and their own. Marietta stopped this cycle because she transcended the normal human reaction and followed Christ's example. She obeyed scripture.

"You have heard that it was said, 'You shall love your neighbor and hate your enemy.' But I say to you, love your enemies, bless those who curse you, do good to those who hate you, and pray for those who spitefully use you and persecute you, that you may be sons of your Father in heaven; for he makes His sun rise on the evil and the good, and sends rain on the just and the unjust. For if you love those who love you what reward have you? Do not even the tax collectors do the same? And if you greet your brethren only, what do you do more than others? Do not even tax collectors do so? Therefore, you shall be perfect, just as your Father in heaven is perfect." Matthew 5:43-48

Marietta prayed for the man who took her daughter's life until she genuinely loved him. God did a work in her heart. She has spoken to many on forgiveness. She has also reached out to the mother of her daughter's killer. Life handed her a hard test, an unbelievably hard test that she passed by forgiving. She did it with God's help. Forgiveness saved herself and others.

Summer Struggles to Forgive

When my parents divorced, I was filled with hatred and rage toward my mother's new husband. I'll call him Roger. I wanted to forgive him, and I tried but I couldn't because the hurt was too deep.

My twin sister Carol knew how I felt and was going through similar feelings, however, because she was away at college. She was being spared some of the agony. Our parents' marriage was our security, and this divorce devastated the both of us.

We were never so happy as we were when our parents married when we were seven. Our new dad loved us enough to adopt us. It meant an end to our chaotic lifestyle with mom always working. We had a stay-at-home mom now. Everything was fine because they were together, we felt secure.

Then Roger hit our family like a train wreck. We were all thrown in separate directions, and we were all wounded. We were never a family again.

Many, many nights I stayed up crying. I didn't want to be filled with hatred and rage and I fought it. Sometimes I thought it was over, but Roger would do some new thing to open the wounds.

My mother moved into our trailer park where my husband and I lived. She bought a trailer right across the street. I dreaded the day that Roger would move in.

I couldn't do it on my own, I couldn't forgive Roger as hard as I tried. But God gave me a miracle. He did it for me.

The day came that Roger moved in, I went over to their house. The feeling in the air was very tense. I felt very awkward. My mother was chattering away about a house they had visited on the way home.

My mother has always been one to exaggerate a story. My sister and I would tease her about it. Today she was going on and on about how beautiful this house that she and Roger had visited was. She was saying "Summer, you wouldn't believe the size of this house and the ceilings, and they had ceiling fans, not just one fan but they had fans."

As she was chattering away about how many fans there were, I heard Roger mutter, "Yeah there were two of them."

It hit me funny. I started to laugh a little, then I laughed, and I laughed. I laughed the rest of the day and as I laughed the anger left, completely. It never returned.

Roger didn't shape up for years, he hurt a lot of people, but I was free of bitterness and anger, it was gone, and it stayed gone. I saw him as a person that needed God, but I never hated him again.

I called my twin sister who was also struggling with the same feelings I was having. While I was telling her the story she started to laugh, and the same thing happened to her. The anger and bitterness left her.

God did for me what I couldn't do myself. Even though I could not forgive Roger, I just kept trying and God honored that effort and finally did it for me.

God wants us free from unforgiveness. He will do that for you too. Is there someone in your life you have struggled with forgiving?

Maybe like Marietta you could take forgiveness one step further and pray for the good of the person that harmed you.

Forgiveness is essential for Christians. It is not easy, but forgiveness sets you free, and forgiveness beats Satan, your real enemy.

Chapter Seventeen
Eternity

Then He spoke a parable to them, saying: "The
ground of a certain rich man yielded plentifully. And he
thought within himself, saying, "What should I do since I
have no room to store my crops?' So, he said 'I will do this:
I will pull down my barns and build greater, and there I will
store all my crops and my goods. And I will say to my soul,
Soul you have many goods laid up for many years; take
your ease; eat, drink, and be merry.'

But God said to him, 'you fool! This night your soul
will be required of you; then who's will those things be
which you have provided?' So, is he who lays up treasure
for himself, and is not rich toward God?"

Luke 16: 16-20

It amazes me how little planning people do for
eternity. People plan for vacation; people plan for
retirement, but they do not plan for eternity. We should
be living each day with eternity in mind! We are not

promised tomorrow. Like the man Jesus was talking about in this story, we need to plan for eternity first of all.

I have worked with older people in nursing homes and home health care for many years throughout my life. I have seen many people die. Some are prepared for it and others are not. I have seen people in their nineties, hours from death and they can't figure out what is happening to them. They have asked me, "When am I going to get better?" Even at that age that can't fathom that the end has come. People walk around with blinders on.

But even if our gospel is veiled, it is veiled to those who are perishing, whose minds the god of this age has blinded, who do not believe, lest the light of the gospel of the glory of Christ, who is the image of God, should shine on them. 2 Corinthians 4:3-4

Satan wants to keep people blinded from eternal things; he does not want them to prepare for eternity. He wants to steal their souls.

What is More Important?

How important is eternity? I recently heard a pastor preach on the subject. He took a string and stretched it across the church and then he took a paper clip and put it on the beginning of the string. "This paper clip represents the length of your life and the rest of the string eternity." Then he asked, "Why do we focus on this little paper clip and miss the whole string?"

Actually, the string could go on forever, but he was

trying to give us perspective on eternity. It really hit home. We put all too much importance on that little paper clip of time and so little on the eternal that never ends.

Think of every person in history that we ever heard about or read about and those also we never heard of, all of them, they all still exist. We think of them as gone or not real, like story book characters that don't really exist. They are real and they do exist. They are in eternity right now, just as we will be soon, either in heaven or hell.

A Certain Rich Man

"There was a certain rich man who was clothed in purple and fine linen and fared sumptuously every day. But there was also a certain beggar named Lazarus, full of sores, who was laid at his gate, desiring to be fed with the crumbs which fell from the rich man's table. Moreover, the dogs came and licked his sores.

So, it was the beggar died, and was carried by the angels to Abraham's bosom. The rich man also died and was buried. And being in torment in Hades, he lifted up his eyes and saw Abraham afar off, and Lazarus in his bosom.

Then he cried and said, 'Father Abraham, have mercy on me, and send Lazarus that he may dip the tip of his finger in water and cool my tongue; for I am tormented in this flame.' But Abraham said 'Son, remember that in your lifetime you received your good things, and likewise Lazarus evil things; but now he is comforted, and you are tormented. And besides all this, between us and you there is a great gulf fixed, so that those who want to pass from here to you cannot, nor can those from there pass to us.

Then he said, 'I beg you therefore father, that you would send him to my father's house, for I have five brothers, that he might testify to them, lest they also come to this place of torment.'

Abraham said to him, 'They have Moses and the prophets let them hear them.'
And he said, 'No father Abraham; but if one goes to them from the dead they will repent.'

But he said to him, 'If they do not hear Moses and the prophets, neither will they be persuaded though one rise from the dead.'" Luke 16:19-31

The rich man from our scripture, who wouldn't even feed Lazarus the crumbs from his table, he is still in torment, thousands of years later and Lazarus is still in heaven.

What could possibly be worth losing your soul? No one in their right mind would allow their soul to be lost! Satan's goal is to lead people to hell, and he uses lies and deception to do it. And when Satan successfully through deception takes a soul from God, he torments them unmercifully.

I had a relative, which as a young man, was offended at church during an offering. Because of this for his entire life he would have nothing at all to do with God. He shut that door for good.

What a stupid reason to go to hell for all of eternity!

I have even heard reasons that are more stupid

than that. "The people in church are all hypocrites; I would rather go to hell than go to heaven with hypocrites."

Or have you ever heard older people say," I have never bothered God all these years. I can't call on God now."

Oh yes you can, and you had better!

Heaven is more wonderful than we could possibly imagine, and hell is more horrible than we can possibly imagine. Jesus gave us in this story of the rich man and Lazarus, a glimpse of hell. There are many testimonies all through history of those who have seen hell. There are many stories of those on their death bed that have seen hell, of seeing devils and demons coming for them after living a life without God. Recently there have been books written of detailed visits to hell. One such book is *Divine Revelation of Hell* by Mary Kate Baxter.

A Modern Woman Visits Hell with Jesus

Mary Kate Baxter was taken to hell over and over for many nights so she could write a book to warn people of hell. The things she saw were terrifying.

She saw people in small pits of burning fire. They looked like skeletons with most their flesh burned off and she would not know if they were male or female until she heard their voices. She also saw large worms that tormented them crawling through their bones. Their captors were horrible looking demonic beings which showed no mercy whatsoever.

As she walked with Jesus through these pits, these souls would beg Jesus for another chance. They all wanted to repent even though it was too late. She spoke to some of the souls in hell who wasted their lives on foolish things. Here is a conversation she heard between a lost woman and Jesus.

"My soul is in torment there is no way out. I know that I wanted the world instead of You, Lord I wanted riches and fame and fortune, and I got it. I could buy anything I wanted. I was my own boss. I was the prettiest, best dressed woman of my time, and I had riches, fame and fortune, but I found I could not take them with me in death. Oh Lord, hell is horrible. I have no rest day and night. I am always in pain and torment. Help me, Lord" she cried. "I planned to serve you someday when I got ready. I thought you would always be there for me. All my life God drew me near to Him with cords of love and I thought I could use Him like everyone else."

This woman ignored her soul. Mary saw many souls just like her who for various reasons lost their souls.

One young man, only twenty-three died in a drunken driving car crash. He wasted his soul on carelessness and strong drink. She saw preachers in hell that did not preach the truth, so that they would be popular with men. And some of the most tormented souls in hell were those who had served Satan, witches and Satanists. The devil lied to them and told them he would keep them from dying and that he would give them a kingdom. Of course, it was a lie, and they were some of those in the worst torments.

Jesus told Mary that Satan deceives people to get them to hell, but God loves and forgives. No one has to go to hell; God is calling each of us to Himself.

This isn't a pleasant subject. After I read Mary's book I felt in shock. I literally trembled the whole night after I read it. It was too much for me and when a friend asked to borrow the book, I told her to keep it. It was years before I dared to pick it up and read it again.

Mary was also sick after her visits and tormented, so awful were the sights she saw.

Another Witness of Hell

Another detailed account of hell is a book by Dr. Roger Mills, it is called *While Out of My Body I Saw God Hell and the Living Dead.* Roger was also taken on a tour of hell by Jesus and showed many things. What he saw was similar to what Mary saw; he saw souls in pits of fire that looked like skeletons. One man he saw was a priest. I will quote the book.

I knew the skeleton was male, even though there was hardly any flesh on his bones. I knew because he began to speak with a man's voice. The skeleton was able to climb completely out of the pit and as he stood in front of me, I was able to examine him from head to toe.

He began to talk and as he did, I watched as his bottom lip fell off to the ground. He said to me, "You can help me! Please help me! You know Jesus; you know

Jesus, don't you? You are a preacher! I have been down here long enough. I thirst all the time, but there is no water! Please give me a cup of water! I am in pain! Great is my sorrow! Please tell Jesus to let me out of this place! I am a preacher too, and I am in here because I was preaching hate, and I was prejudiced. I used to do pastoral work at a church in downtown Detroit, named St. Mary's Catholic Church. I was a practicing priest during the 1960's. I was there during the 60's riots, and I did not like any of the black parishioners that walked through the church door. I was racist, and in my opinion, I thought I was doing God a favor by hating and being prejudiced toward anyone that was not of my race. I was raised as a Caucasian Catholic, so I thought it was the righteous thing to do as far as disliking certain nationalities. That was my opinion while I was alive on earth, but when I died, demonic spirits brought me here and threw me in this pit. It was then I was told the reason why I am here. I was shocked and surprised to be here in hell! I began to pray to God and say, 'I preached for You for many years! I worked in your church! I did pastoral work! Why am I here?'

Then God appeared to me here in hell, and He told me that I was here because I judged my black brothers and sisters and that I had hate in my heart."

There is more to the story, and I encourage you to read the book, but I want you to see how this man was deceived by Satan. He thought that even though he was a priest he could disobey God's word and hate others.

If the Bible says something is wrong, it IS wrong

and if it says something is right it IS right. This man believed he was serving God by spreading hate and violence. People cannot hide their sin in eternity like they can on earth.

Roger saw many people in hell; another was a young man who was only twenty. He joined a gang and used drugs. His parents tried to warn him, but he wouldn't listen. While on a street corner with his gang, a rival gang shot him and killed him, he was taken to hell.

I don't like to talk or think about hell, but I want to expose Satan and make several things very real on these pages. Several things that Satan blinds people's minds from and they refuse to think about they are:

How long eternity is {Forever and ever and ever}.

Hell is real and it is horrible.

The word of God is true, and it has to be obeyed {many churches no longer preach the true word of God, many denominations embrace openly abominable sins.}

There is no reason to spend eternity in hell and nothing more important than living for God and spending eternity in Heaven with Him.

Think about it every day and let it guide you in the decisions you make, how you spend your time and how you treat people.

I want to list some of the stupid reasons that some people have let themselves spend eternity in hell. {I bet you can add some more to my list}

They are turned off by a Christian or hurt by them.

They blame God for a horrible event that happens in their life.

They think they have more time and put off serving God until later. {No one is promised tomorrow}

The cares of this world, the temporal things like houses cars, wealth seem more important than serving God.

Sin, such as sex sins, our society is obsessed with sex and perversion. {Turn on the television}

Vanity, outer beauty is a stupid reason to lose your soul. The Bible likens us to flowers that fade. We need to be more concerned with our character. Women ore obsessed with youth and shape, risking their lives on surgeries to make them more desirable, some have even died during those surgeries.

Alcohol and drugs, I personally don't think it is okay for a Christian to drink and I have heard every argument such as Jesus turned water into wine. There are many souls in hell because of alcohol and that is why I won't touch alcohol. If you are addicted to drugs and alcohol, don't stop trying to get free, if you don't give up you WILL make it!

Carelessness, many people live good lives but leave God out of them, we have no hope without God.

Desire for fame, look at what people do on screen for money. I saw an interview on television of a handsome young actor from a cop show. They were interviewing him because his character had to do nude sex scenes for the show this season. In real life this young man had recently married, and his new wife was expecting a baby. He stated on the show "I feel like I am cheating on my wife." Then he added "But this is my career, and I am an actor."

"Don't do it!" I screamed at the TV, "You are cheating on your wife, and it isn't worth it!!!!"

False religions and the occult, the ultimate deception, Satan wants you to worship him through false religions.

Unforgiveness, anger and hatred, we are commanded to love and forgive.

There are many more reasons people go to hell. None of them are worth it.

As awful as hell is, Heaven is wonderful beyond imagination. There are so many testimonies of people seeing Heaven, one of my favorites is Jessie Duplantis and he has a book and a video. My kids were still pretty young when I bought Jessie's video of his visit to heaven. They would come home from school and every day they would watch that video, over and over. My daughter Lonna said to me "I can go through anything during the day because I know when I get home, I am going to watch the video about heaven."

Heaven

Heaven is too good to be true, but it is true. Whereas those who go to hell describe the horrible stench, the awful sounds, the heat the thirst and the pain, those in Heaven describe wonderful scents, beautiful sights of flowers and scenery with colors beyond belief, beautiful music everywhere and delightful things to eat.

Every desire is met in Heaven. When Jessie saw his

home, it was decorated exactly to his taste with the things he loved. His home was lovingly planned for him just the way he liked things. Those who love horses, have horses. Those who love to garden have beautiful gardens. The cooks have beautiful kitchens to prepare food in. Jessie met Abraham and Jonah, Paul and King David. Everyone is family.

But the greatest place in the whole universe is the throne of God. The joy around the throne of God is incomparable. Millions gather there from every tribe and nation to worship God. Jessie could hardly lift his head in God's throne room the glory of God was so overpowering. This is where God wants you to spend eternity. I think we need to think about heaven every day. Jesus said we could store up treasures there. [Matt 6:19-21} We need to live like heaven is real, because it is!

Warn those on Earth

One thing I find interesting, in just about every testimony of Hell, even the one Jesus gave us of the rich man and Lazarus, is that the lost souls in hell want to warn those on earth. The rich man wanted to warn his five brothers. Roger Mills encountered this also when he went to hell. Remember the priest he was talking to. I will quote it for you.

He told me his name, he told me where he lived, he told me what his house address was, he recited his social security number and he recited to me his

telephone number. Then he said to me, "Will you ask God to let me out of here? If he does let me out of here, tell Him I promise to go back to the earth and apologize to every black person I ever offended, those that I know and those whom I will meet! If He doesn't let me out, would you go back to earth and tell my two sisters that I died and went to hell, and warn them not to come to this horrible place."

Roger did find his two sisters and told them the message from their brother. One hung up the phone on him and the other believed him. Roger talked to another young man who recognized him from earth. The young man in hell, called to Roger and said to him,

"You know Tony, don't you? I know him. He is friends with my relatives. Tell him to go to my cousins and uncles and tell them I am in hell, and not to come to this place. The boy told me, "Tony knows me, I use to hang around in his neighborhood; I used to walk up and down the streets; one street in particular- Hendricks in Warren Michigan. I did not want to come to this place. I did not believe in its existence. I hung out with many people who sold drugs and used drugs. I got into a lot of trouble. People that I trusted turned me in to the police for dealing with drugs, but that is all I knew. I lived pretty much a rough life. I hardly ever went to church. I went when I was very young on Christmas and Easter, if then. No one of my family was of any Christian influence. I committed suicide indirectly. I never meant to kill myself. I had alcohol in my system; I had drunk plenty of beer that evening, right up to the moment I died. I sniffed

155

cocaine; too much of it I would suppose. I wanted to vomit. I felt very ill, so I said to my girl, 'I am going to the bathroom. I'll be back in a minute.' There were about fifteen people there at the house. I can remember going to the bathroom and leaning over the toilet, where I fell on my knees. It seemed I got lightheaded, and I remember falling to the floor, and I came here. I am in torment and I am frightened all the time. If only I could go back to earth, I would tell all my family and friends not to come here."

Eternity is forever. Hell is horrible. The Bible is true, we need to believe it and obey it. There is no reason good enough to lose your soul. Satan is a liar and a deceiver; he lures people into hell.

The people in hell right now wish they could tell you something. If you could hear them, they would tell you this, "Don't come to this place!"

CHAPTER EIGHTEEN
Psalms Ninety-One

He who dwells in the secret place of the Most High
> *Shall abide under the shadow of the Almighty.*
> *I will say of the Lord, "He is my refuge and my*
fortress my God in Him will I trust."
> *Surely, He will deliver you from the snare of the*
fowler
> *And from the perilous pestilence.*
> *He shall cover you with His feathers,*
> *And under His wings you shall take refuge;*
> *His truth shall be your shield and Buckler.*
> *You shall not be afraid of the terror by night,*
> *Nor of the arrow that flies by day,*
> *Nor of the pestilence that walks in darkness,*
> *Nor of the destruction that lays waste at noonday.*
> *A thousand may fall at your side,*
> *And ten thousand at your right hand;*
> *But it shall not come near you.*
> *Only with your eyes shall you look,*
> *And see the reward of the wicked.*
> *Because you have made the Lord,*
> *who is my refuge,*
> *Even the Most High, your habitation,*

No evil shall befall you,
Nor shall any plague come near your dwelling;
For He shall give His angels charge over you,
To keep you in all your ways.
They shall bear you up in their hands,
Lest you dash your foot against a stone.
You shall tread upon the Lion and the cobra,
The young lion and the serpent you shall trample
underfoot.

Because he has set his love upon Me, therefore I
shall deliver him;
I will set him on high because he has known my
name.

He shall call upon Me and I will answer him;
I will be with him in trouble;
I will deliver him and honor him.
With long life shall I satisfy him,
And show him my salvation.

I want you to have this psalm with you always. I want you to take time to memorize it. It will be like having a powerful weapon you can pull out when you need it. I have used it many times. One time my daughter, Joy, called me in the middle of the night.

"I need you to help me pray, "she said. "I feel such evil in my house, and I can't seem to pray it away." She had been going through much chaos in her life. Now she had woken up in the night and literally felt an evil presence in her home.

"I will be right over," I said. I jumped in my car still

in my pajamas and rushed over to help her pray. When I got there my hair literally wanted to stand on end. I could feel the oppression. We prayed together for quite some time, but the heaviness still filled the room. Then I began to quote Psalm 91. I could feel the power of God as I said each word. The atmosphere changed and we finally felt the heaviness lift.

Joy told me that she saw Heaven as I was quoting scripture. She saw us both come into the throne room of God. We were battered and torn from battle. The throne room was filled with a myriad of worshippers dressed in white but as we came in the sea of people parted for us and allowed us to go to the front and stand be for the throne. As we entered and stood before the Father, we were granted favor and angels were dispatched for us. Quoting Psalm 91 powerfully changed the atmosphere and peace was restored to my daughter's home.

Many soldiers have learned this powerful scripture and used it in battles, for protection.

I want to keep this chapter short because my purpose is for you to take some time and learn it. It will then be yours to use whenever you need it.

Arm yourself!

Chapter Nineteen
What Lies Have You Believed?

For the weapons or our warfare are not carnal but mighty in God for pulling down strongholds, casting down arguments and every high thing that exalts itself against the knowledge of God, bringing every thought into captivity to the obedience of Christ.   2 Corinthians 10:4-5

If there is anything in your belief system that is contrary to the word of God, you have swallowed a lie from the enemy. He sets up strongholds in our minds based on lies he can get us to believe. These strongholds in our minds actually get us to believe opposite to the word of God.

God is Not Angry with You

One of the biggest lies Satan tells people is that God is angry with them or that God will not forgive them. God is not angry with you. Does He like sin? No, He doesn't. Does He punish sin? Yes, He does. But if you repent of your sin, He forgives you completely.

Why would Jesus come to earth, be tempted by the devil and go through a whole life on earth and then be beaten and crucified, so that He could atone for your sin, and then not forgive you? If His plan was not to forgive you, He could have saved Himself a lot of trouble! His plan is forgiveness and redemption, and He paid a humongous price for it. You are that precious to Him.

If you had a loved one in jail, would you pay a one-million-dollar bond for them and then let them sit in jail and not get them out? No if you paid their bond, you would get them out of jail. Jesus did more than pay your bond; He did your prison sentence for you!

We are Not Free to Sin

And then there are those who swing to a lie on the other end of the spectrum. "I can sin," they tell themselves "And Jesus will forgive me." So, they deliberately treat the most precious ransom paid for them in blood, in utter disrespect. They may wake up in hell.

There is a big difference though between those who are trapped in sin and those who willfully sin. God is very patient with those who are trying, He sees our hearts.

Satan has a whole bag of lies he uses on humankind. We all have lies we believe even now, strongholds in our minds that go against the word of God.

I remember reading a story in one of Kenneth Hagin's books about a woman who believed a lie. Brother Hagin was praying for this woman; she was having migraine headaches. She had them often and had been suffering with them for years. As he was praying for her the Lord showed him why.

Years before when she had first begun to serve the Lord, she had told a lie. The devil convinced her she had committed the unpardonable sin. She was suffering because she was sure she was going to hell when she died. The lie was causing her migraines. When the Lord revealed this to Kenneth and he exposed the lie from the devil to the woman, her migraines stopped. Satan was able to oppress her with headaches for years because he got her to believe a lie.

Summer Falls for a Lie

I have had experiences where the devil has convinced me of a lie. One such lie concerned my children. I was walking in faith when it came to my children and their protection. And God was always coming through for me. I can tell you of so many times God helped me, and kept them safe, so I should not have fallen for a lie.

Such as the time when my husband, Jim and my son, Jamie who was three and a half at the time, went out wandering on a Sunday afternoon. After they had been gone awhile, I started to feel very agitated. I stopped everything and I began to pray for them. I prayed until the agitation lifted. I wondered what had happened.

Soon my husband came home. He was white as a sheet, and he had quite a story to tell me. My husband, a horse lover, decided to find a horse farm to take Jamie to and let him see the horses. They found one and were enjoying a tour by one of the workers. As she was leading them into a large barn with double doors, suddenly the doors flew open. A charging horse that had gotten loose, burst through. Jim said the doors pushed him one way and Jamie the other. Jim could see Jamie was right in the way of the charging horse, but Jim couldn't get to Jamie. Jim said he literally saw the horse run over the top of Jamie.

Jim was horrified! He ran to Jamie, expecting the worst, and picked him up. It was the worst thing imaginable for a parent seeing his son crushed by a charging horse! But to Jim's absolute amazement Jamie was fine, not a scratch not a bruise, nothing! He came home shaking but praising God. We knew it was a miracle, one of many. God had come through for us so many other times too. Like when our daughter Lonna drank the dry gas in chapter 7. So, I shouldn't have fallen for Satan's lie.

My trouble started when I heard a little boy at our church died. I knew who he was. I had seen him in Sunday school. The family had three darling little boys and he was the middle one. He was only about six years old. I'd heard that he had been throwing up and his mother brought him to the emergency room, but he died the same day.

I couldn't believe it when I heard about it. I was so devastated for her. I tried to express to her my sympathy to her when I saw her, but I couldn't even speak. She almost comforted me; she said God had given her peace. But I didn't have peace, I thought, "How could that happen to a Christian?"

Then came the lie, "God didn't protect her little

boy; He is not going to protect your children either."

I began to be filled with fear. Satan had hooked me with his lie. Now, the fear left an open door for the enemy and soon, my daughter, Lonna, who was two and a half, became sick.

We had just moved to Florida and things were very different there. In Michigan I would see a doctor and they would let me pay when I could and usually it was under twenty dollars. We got to Florida and the doctors wanted fifty dollars up front. It might as well have been a thousand; I didn't have fifty dollars. I called every doctor in the phone book, no one would see her. {Don't worry God was in this, He was dealing with the lie}. I was filled with fear, she was running a fever and I couldn't get it down. Ten fear filled days went by and she was still running a fever. I felt so helpless.

I left her home with her dad and went to a good, faith filled prayer meeting for prayer. The ladies all gathered around me to pray. God revealed the problem to the ladies that were praying. I had believed a lie about God. I thought that He wasn't caring for the little boy who died, and I was full of fear that Lonna would die too. It was a lie.

I had to deal with the lie. I put the life of the little boy, who died, in God's hands where it belonged. In the God who loved and created him and knew what was best for him, better than I. I didn't have to know the reason. I just believed that God knew best, and the child was with Him. The lie was dealt with, I was trusting God again and the fear was gone. When I got home Lonna was fine, the fever had finally broken.

What Lies Have I Believed?

I have learned I can't get my eyes off God's word and my own life. I am not going to understand everything that happens to other people, like why that little boy died. I don't know all the pieces to the story, maybe God was protecting the child by taking him home with Him. I don't know. If I don't understand something, then I need to leave it in God's hands knowing that GOD IS GOOD! AND HIS WORD IS TRUE! AND I CAN ABSOLUTELY TRUST HIM! That is truth!!!!

I asked God one day "What lies have I believed?" He gave me a whole list.

That you have no future.

That you won't fulfill your destiny or live up to what I have called you for. You think that I will do My part and you will blow your part.

That you need to hide from Me. {That is Summer hiding from God}

That it doesn't do any good to pray. {What a lie, pray and don't stop}

That every day is the same you just exist and go to work, there is no purpose.

That if you lose your job, you will lose everything.

That I, [God} am far away.

That if you get too close to the Holy Spirit, you will blaspheme Him.

What a list! We need to bring the lies the devil brings against us down, bringing them into the submission of the knowledge of Jesus Christ.

What lies have you believed?

You know what to do, believe God instead.

CHAPTER 20
Follow Jesus 'Example

If we allow the devil to, he will try to destroy us. He will put lies in our minds to deceive us; he will use people to hurt you and try to fill you with unforgiveness and bitterness therefore giving him access. He will lie to you about God and try to get you to believe God doesn't love you so you will give up. Or he will swing to the other extreme and convince you to sin willfully and tells you Jesus will forgive you, so you can sin, it doesn't matter. He wants to destroy you, and he has many tricks. What are we to do????

Jesus is our example. Jesus is our example. Jesus is our example.

He committed no sin neither was any guile found in His mouth, who when He was reviled did not revile in return; when He suffered, he did not threaten but committed Himself to Him who judges righteously. 1 Peter 2:21* In other words, Jesus put things in God's hands! He gave the situation to God.

All through the gospels Jesus tells us how to defeat the devil. He tells us to love our enemies, to forgive, not to repay evil for evil. It is time for us to really put these things into practice. We believe them but do we do them? I know I fall way short. But this is how we defeat the enemy. We can't afford to get into a struggle with another person. Satan's plan is to attack us through someone else, unleashing his evil against us. Then if he can get us to fall for the trap and retaliate in some way the evil is multiplied. If we respond how Jesus tells us to, with love, the evil is vanquished.

Another Battle

I had a real battle with the enemy many years ago, but Jesus helped me to come through it. I got into trouble, but I was just trying to be nice to someone.

I met a lady at church who said she needed a place to stay. She seemed so nice. My husband was not home. He was in an alcohol rehab facility for a couple of months so I thought I would help her.

Her name was Elsie. I told Elsie she could stay with me and my three children for a month or so, while my husband was gone. I thought she could work and save money by staying with me, so she could get a place of her own.

Please learn an important lesson from me, pray before you ever let someone come and stay with you! I thought I was doing something good, which would please the Lord.

I went to pick up Elsie where she was staying. She was staying with a group of single Christian ladies from our church. They rented a place together. I couldn't figure out why they wanted her to leave, she seemed so nice.

The leader of my prayer meeting from church called me to warn me. She told me Elsie operated out of a spirit of witchcraft, and these ladies had a really bad time with her. I did not even know what that meant, operated out of a spirit of witchcraft, wasn't she a Christian? I didn't know what to do, I had already let her move in.

I soon found out I had made a big mistake. The first trouble I had was the first night. Elsie put her bed up in my living room. It was the only place to put it, we only had two bedrooms, a bath and a kitchen and the living room. I worked doing a paper route and had to leave about two a.m. To get out the front door I had to walk through my living room.

Elsie yelled at me that I couldn't come through the living room at night. She didn't want to be disturbed. That was just the beginning. She brought a cat with her. The cat pooped on my furniture; Elsie refused to clean it up. Elsie dominated the house.

And forget about her saving her money to get her own place, she didn't work. Elsie spent hours every day putting on make-up and fixing her hair, then getting dressed. Her hair was dyed bright red, and she wore bright red lipstick. She looked very nice when she was done but very painted. Then she would go to the church and prey on the body of Christ.

She got the Christian school at our church to let her direct the Christmas program. It was not done with the correct spirit. It was not pleasant for the kids; she was too hard on them; it was done to bring attention to herself.

Another thing Elsie did was pick on my children. My seven-year-old daughter, Lonna, was her primary target. She did not seem to like children.

I did not engage in open conflict with Elsie, but I was very upset, and I did not know what to do. Even though Elsie was supposed to be a Christian, she was a witch. I had a witch living in my house.

Witchcraft has to do with control. The devil had gotten a foothold in my house through Elsie and my home was under attack.

My husband ended up coming home from rehab early and found out I had let Elsie move in. My husband is a tough guy. He has lived a rough life. He put up with Elsie for one day. As soon as she left the house, he pulled her bed and all her things out the door and put them on the driveway. Then he locked the door.

She came home and I hoped she would leave but not so. She talked my mother into letting her move in with her. We lived in a duplex and my mother was next door. Elsie was still close.

One day my daughter Lonna was next door playing when Elsie started to yell at her and chase her. I heard the commotion and opened the door between my apartment and my mother's. Lonna ran through the door and Elsie was right behind her. When Lonna ran through the door I stood in front of the door and blocked it so Elsie couldn't come in. Elsie started yelling at me that Lonna needed a spanking and tried to get into my house.

I was past angry I was filled with rage. I don't like confrontation, but this was unavoidable. As Elsie tried to push past me, I literally pushed Elsie back through the doorway and closed my door and locked it. I stood there heart pounding, I did not know how to handle being so angry.

There was a spiritual warfare going on. Elsie was a tool of the devil. She was spinning a web of trouble all around us, it was spiritual, and it was demonic.

I began having a bad dream. Elsie and I were at the door, just as had actually happened, but in the dream, I couldn't shut the door. Her arm was in the door reaching out at us. I did not know how to fight this spiritual battle.

I did a lot of praying and the Lord showed me what to do. I had to let go of all anger and frustration to avoid Satan's trap. I told the Lord I didn't think I could do that. He told me how. "I want you to pick one good thing about her and dwell on that,"

It was hard for me to come up with something positive about Elsie. But I thought about it. Jim and I had been having trouble since he came back from rehab. Elsie had made a comment to me that I should have Joyce counsel us.

Joyce was a lady from church who was a counselor. We did have Joyce counsel us, and it was a good experience. That was actually the only positive thing I could think of about Elsie, but it worked. Instead of allowing my mind to dwell on all the things that were making me angry about Elsie, every time she came to mind, I thought, "she got us to go see Joyce."

It worked. It worked wonderfully. I felt no more bitterness or anger, it freed me to see Elsie in a different light.

Elsie was being used by Satan, but Elsie was also bound by Satan. If I had stayed bitter and angry, the witchcraft spirit operating through her would have defeated me. The Lord in His wisdom showed me the way through this evil web that surrounded me. Had I not responded correctly, evil would have been multiplied, instead it was vanquished.

Not only did I become free from the bitterness and anger, but Elsie also left my mother's house, and I did not have to see her anymore. I believe this was because I defeated the spirit working through her.

We have to respond in a spiritual way. I could not have done this by myself. I was too angry, especially because she was attacking my daughter. That is one of my anger buttons and she pressed it. I don't like people messing with my kids.

Jesus stayed free in his life on earth. He teaches us how to stay free. We are to love and pray and do good to our enemies. It is all written in the Sermon on the Mount in Matthew chapter 5.

Jesus also shows us in how He lives His life. He obeys God, He follows the Holy Spirit, He walks in wisdom and love. Jesus on the night before His death tells His disciples, *"I will no longer talk much with you, for the ruler of this world is coming, and he has nothing in Me."* John 14:30

Satan had nothing in Jesus. Jesus did not engage in human conflict. He fights His battles differently. He loves his enemies, He vanquishes evil.

Jesus lived a perfect life. He is our example. When we walk with Jesus as best as we can, like Jesus, through faith in Him. Then we are covered by His blood, and we have His righteousness. We can live like Jesus did, overcoming the enemy, even though we aren't perfect. Did Jesus face trouble? Yes, but He defeated the enemy every time! Remember what the Bible tells us, *He did not revile in return but committed his soul to him who judges righteously.* I want to see you begin to defeat the enemy in your life! No, you don't have the plan God does. Your job is to commit things to Him. We have to do this a lot.

Last year I tried to refinance my home. I had been struggling to get the bills paid and I thought if I got the house payment lowered it would help. Also, I would get to skip a house payment and I could pay some looming bills that I had no money to pay.

I called a bank, and they told me I could lower my payment by eighty-five dollars per month. We talked it over and my house would have to appraise for at least 155,000. I thought it was no problem because my last appraisal at the height of the housing crisis was 163,000. The housing market had improved since then and I had made improvements on my house since then.

When they asked for a five-hundred-dollar deposit, that would go toward closing costs but was nonrefundable, I did it thinking there would be no problem. I put it on a credit card. {I didn't have five hundred dollars, my finances were a mess, because I had had two surgeries a couple months before.}

A woman appraiser came to my house. She appraised my house at 139,000. I was shocked and angry. I knew my house was worth more than that just because of the size. We have seven bedrooms and three full baths, and our house looks over a lake.

The devil knows the way to get me is money. I have never made a lot and money to me means hard work and overtime. I was mad. We called the appraiser to try to reason with her and she wouldn't budge.

Now my finances were even more dire. I just wasted five hundred dollars. I asked my business smart dad what I could do about it and he said I could post a complaint on a consumer website so others wouldn't use her.

I wanted to because I was angry. I wanted to get even. But I didn't. I thought, this stuff I write about, I need to practice it. So, I let go of the anger and I put it in God's hands. Every once in a while, when I would look at my finances, the devil would stir me up and I would get mad at myself for wasting five hundred dollars and I would have to give it to God again.

God did a miracle. It was His plans not mine. Six months later I was still having trouble financially. So, I went to a different bank to ask for a loan. They couldn't give me a loan, but he said, "We can refinance your house."

I said, "No I don't think the appraisal will come in high enough." He told me he didn't need an appraisal. To make a long story short, they refinanced my house. They gave me a five-hundred-dollar grant toward closing costs. {I'd never heard of such a thing.} They lowered my payments over one hundred dollars per month and I was able to skip two mortgage payments which enabled me to catch up with all my bills.

I get amazed at God's faithfulness. When we follow Jesus example it defeats the devil! I defeated the devil in this situation by following Jesus' example. I put the situation back in God's hand.

I could give you many more examples also. Satan wants to get us in strife so he can destroy us further. God wants us to cast our cares on Him and stay in peace so He can bless us. God wants us to watch our words, forgive, stay in peace, and commit ourselves to Him.

Jesus wants us to have life more abundantly, that is His plan. We have switched kingdoms. We belong to Jesus now. Satan has no power or authority over us anymore. When we follow Jesus' example, we defeat the enemy as He did.

CHAPTER 21

You Belong to Jesus

Dear Friends,

There was a time I lived in deep darkness. Because of the emptiness inside me I had embraced darkness. I know I wouldn't have lived much longer if I had continued down the path I was going. I was truly miserable. Satan had me and he knew it. I was bound in darkness. Satan thought he had me forever, that he had my doom all sewed up. I couldn't even stand hearing about Jesus; it made me cringe.

But Jesus carefully planned my rescue. His timing was perfect. He cut off the unholy protection; He reconnected my shame wires and opened my eyes to the ugliness of darkness. Then Jesus came to me personally and told me He loved me.

I entered the kingdom of Light and I never looked back. The difference was unbelievable, like waking up from a terrible nightmare. Six months later was that awful night I heard the voice that told me he was coming to get me and take me to hell. When he came to get me, he was met by Jesus. Of course, I had no idea Jesus was in the room. I did not even know Jesus was stronger than Satan. I had very little teaching.

When Jesus spoke, that thing that had entered the room left immediately.

But Satan was furious. The devil always viciously seeks revenge. That night began torment that lasted for years. Not just torment, every two weeks or so someone I knew, or a family member died. It began with my grandmother that same night. Altogether six people died.

I had begun a battle that was to last for ten years, although there were victories along the way. One of the greatest victories is when God finally broke the lie that had me bound.

Remember in the introduction of this book when I talked about my sister, Carol, speaking a word from the Lord to me, in a booming voice I hardly recognized. She had come over to pray that night. We were trying to pray but it felt like our prayers were bouncing off the ceiling. It felt heavy. We kept persevering. A tiny thought, like a little bubble, crossed my mind; it was such a faint little thought I almost missed it. "Bind the hindering spirit" it said softly.

I thought, "What could it hurt?"

"I bind the hindering spirit," I added as we were praying. That is when the atmosphere changed. Suddenly I felt as if I were seated before God. My sister's voice changed to a booming voice that sounded like thunder.

"Satan has no power over you! He has no power over you, no power over you!" God's voice through my sister roared.

I didn't believe it, at least not at first. "He does" my mind answered. "He torments me."

"Satan has no power over you, he has no power over you he has NO power over you, no power."

My mind was having a struggle, I know God doesn't lie. It must be true, but Satan had been tormenting me for years, my nights were pure terror. I tried not to anger him so it wouldn't get worse.

God continued, "Satan has No power over you, no power over you, No power." His words continued on and on. Each time He said it, the lies in my mind took another blow, His words were having an impact.

Finally, I believed Him.

"Satan has no power over me."

I still had a battle to win before the oppression stopped, but I had the truth now. Satan has no power over me, I belong to Jesus. It was not too much longer that I received victory and all the torment stopped.

I had learned a few things along the way. Satan tormented me yes, but it was only torment. He could not have me, he could not get me, he tried. Jesus personally stopped him. Also, I learned I had access before the throne of God. That is when Jesus threw Himself before the

Father on my behalf. And I also learned that I had all the resources of heaven behind me. And then later I even found out that when I respond correctly to his attacks, by following my example, Jesus, I get back everything he has stolen and even more. I defeat him.

Over the years I learned even more. I fell for Satan's deception a few times and I learned if God tells me something to stick to it, not to listen to reason, just obey.

I learned that to stay in peace keeps me safe from the enemy, but he will attack that peace by pushing my buttons through other people. I had to let go of the frustration and anger and get back to peace.

I learned how important it is to follow the Holy Spirit. I learned about the ultimate weapon, love and its great ability to deliver those in darkness. And don't forget I also learned not to give Satan permission with my mouth. I learned the many lessons I have taught you in this book.

I want you to begin to defeat the enemy in your life! I hope this book has helped you to do that. Remember, you belong to Jesus!

I belong to Jesus; those words mean so much to me. The devil came to destroy me; he told me he was coming to take me to Hell. I believed him, I thought I was doomed. But unbeknown to me I wasn't alone. Jesus was standing next to me. I heard Him rebuke the devil with a voice that boomed with authority, "Leave her alone! She belongs to ME!"

You see, I had switched kingdoms. I had been translated out of the kingdom of darkness and now I belonged to the Light! Do you realize what that means? It means;

Satan has no power over me, I belong to Jesus now!

God bless you and much, much love from,

your sister in Christ,

Summer

Epilogue

This epilogue is for anyone who does not yet know Jesus. You too can be translated or zapped out of the kingdom of darkness and into the kingdom of light. You can belong to Jesus. Jesus has paid, with His own blood, for your salvation. Your part is to ask Him to forgive you and to give yourself to Him. I was so desperate when I came to Jesus. I was miserable. Jesus told me He loved me and from that moment on my life has been about Him. I am so glad I belong to Jesus! Jesus loves you; would you like to belong to Him?   Pray with me.

Dear Jesus, please, forgive my sins. Take all of me. I want to live for You.

If you prayed, now say it. I belong to Jesus! I belong to Jesus! I belong to Jesus!

Notes

Chapter 6   *Tormented Eight Years and Back*
By Peggy Joyce Ruth Impact Christian Books Inc.
Kirkwood, MO. 63122

Chapter 11   *The Priestly Bride* By Anna Rountree
Charisma House, Lake Mary, FL.32746

Chapter *13*   *Run Baby Run*
by Nicky Cruz with Jamie Buckingham
Bridge Publishing Co. S. Plainfield, New Jersey
pages 104-105

Chapter 14   *Spiritual Warfare; The story of Stephan Morin*
CS 663
Focus on the Family, Colorado Springs, CO. 80995

Chapter 16   *The Lost Child* by Marietta Jaegar
Zondervan

Chapter 17   *A Divine Revelation of Hell* by Mary K.
Baxter   Whitaker House, New Kensington, PA.
                                    Chapter 2

While Out of My Body I Saw God Hell and the Living Dead
                        by Dr. Roger Mills
Trinity Publisher, St Clare Shores, MI. 48080-6471
                        pages 39-41 and 45-46

Close Encounters of the God Kind by Dr. Jesse Duplantis

All scripture references are New King James Version
                        Thomas Nelson Publisher